T0175833

Bots and Beasts

Bots and Beasts

What Makes Machines, Animals, and People Smart?

Paul Thagard

The MIT Press
Cambridge, Massachusetts
London, England

© 2021 The Massachusetts Institute of Technology

All rights reserved. No part of this book may be reproduced in any form by any electronic or mechanical means (including photocopying, recording, or information storage and retrieval) without permission in writing from the publisher.

This book was set in Stone Serif and Stone Sans by Westchester Publishing Services. Printed and bound in the United States of America.

Library of Congress Cataloging-in-Publication Data

Names: Thagard, Paul, author.
Title: Bots and beasts : what makes machines, animals, and people smart? / Paul Thagard.
Description: Cambridge, Massachusetts : The MIT Press, [2021] | Includes bibliographical references and index.
Identifiers: LCCN 2020040791 | ISBN 9780262045940 (hardcover)
Subjects: LCSH: Intellect. | Animal intelligence. | Computer logic. | Cognitive science.
Classification: LCC BF431 .T423 2021 | DDC 156/.3—dc23
LC record available at https://lccn.loc.gov/2020040791

10 9 8 7 6 5 4 3 2 1

For my smart sons, Dan and Adam

Contents

Preface

Magpies can recognize themselves in mirrors, and octopuses can open jars to get at food. An IBM computer won a debate with people about the costs and benefits of artificial intelligence, and cars are beginning to drive themselves. How do computers and animals measure up to the minds of people?

Human intelligence involves more than IQ, thanks to many features such as decision making, learning, and creating. These features generate report cards for evaluating smart computers and animals. The ability to learn how to learn is just one of the advantages that people have over current bots and beasts. Consciousness only began when animals such as fish developed brains large enough to combine thoughts. Artificial intelligence is advancing but still falls short of the human capacities for consciousness, emotion, and creativity. These conclusions have important implications for the moral treatment of animals and intelligent computers.

In this book, I make the first systematic comparison of intelligence across machines, humans, and other animals. I draw on philosophy for a method of attributing mental capacities to nonhumans and for an approach to ethics based on vital needs. Psychology and neuroscience furnish the mechanisms that support intelligence. The assessment of limitations in approaches to AI today provides a clear blueprint for future research to approximate human-level intelligence. I recommend policies to ensure that such developments promote the satisfaction of human needs.

Additional material and live links for the websites mentioned in this book are available at paulthagard.com.

1 Attributing Minds to Machines and Animals

My friend Laurette has two cats, Zhanna and Pixie, shown in figure 1.1. When Laurette pets Zhanna, Pixie often interferes by attacking Zhanna. By analogy to humans, it is natural to interpret Pixie's behavior as jealousy, but perhaps Pixie is just attempting to assert dominance or establish territory. What mental states can we attribute to such animals? I use the term "animals" to mean "nonhuman animals" and sometimes use "beasts" for the same purpose.

Ascribing mental states to machines generates puzzles similar to those that arise with animal minds: when is such attribution justified? When I announce to my Amazon Echo that I am home, Alexa sometimes responds, "Well, hello. I'm very happy that you're here." Is Alexa actually happy that I'm home, or is the response just a canned phrase prepared by a human? Similarly, are the self-driving cars that are now making their way around various cities really driving themselves, or are they just following commands produced by human programmers? How can we tell if machines have mental states? I use the term "bots" to cover robots and other kinds of intelligence accomplished by computers.

A hundred years ago, humans seemed to be the only intelligent beings on the planet Earth. People established their smarts by developing art, literature, science, mathematics, technology, and complex organizations. In comparison, animals such as cats, dogs, and cows seemed pretty dumb. Machines such as the Ford Model T seemed even dumber.

But consider the New Caledonian crow, a species that lives on an island near Australia. These birds eat food that includes nuts, seeds, and insects, but they do much more than just peck away at whatever looks tasty. To capture grubs hiding in cracks, the crows modify twigs and leaves to turn them into tools that they insert into the cracks. When the grubs bite the

Figure 1.1
Zhanna and Pixie. Photo by Laurette Larocque.

tools, the crows pull them out and devour the grubs. Not only are the crows using tools, but they are using their beaks to make the tools by getting the twigs and leaves into the right shape.

Other beasts are also capable of impressive feats of intelligence. Bees can learn the location of food sources and perform dances that tell other bees where to find them. Octopuses can solve complicated problems, such as removing lids from jars and learning to navigate mazes. Prairie dogs communicate with one another by signals that indicate the size and appearance of interlopers. Elephants display complex emotions, including grief. Dogs can pick up on the emotional states of their owners. Dolphins use clicks to communicate with one another. Whales use sounds to coordinate group activities. A chimpanzee learned 350 signs of American Sign Language and taught some to her son. Bonobos use sex to resolve group conflicts. Members of at least eight species can recognize themselves in mirrors, including orangutans, gorillas, pigs, and magpies.

Machines are also turning out to be much smarter than previously suspected. A hundred years ago, the closest things to intelligent machines were mechanical calculators that used hand cranks to add numbers. But more flexible computers were developed in the 1940s, and in 1956 the field of artificial intelligence was born. Also known as "AI," "machine intelligence," and "computational intelligence," this field has grown spectacularly and is having a huge impact in current technology. Driverless vehicles are being developed by many companies, and people are accustomed to using the effective speech recognition systems used in Amazon's Alexa and Apple's Siri.

Some technologists look forward to a "singularity," when machines will surpass human intelligence. Others are more concerned: Bill Gates, Stephen Hawking, and Elon Musk have issued ominous warnings about the risks to humanity of future dominance by artificial intelligence. Whether these warnings are warranted depends on careful assessment of the current state and long-term prospects for machine intelligence.

Are computers already smarter than cats? Outdoor cats can still find their way around their environments and obtain energy much more effectively than any robot today, but cats fall short of computers that can process language and play complex games. Comparison of the mental strengths of bots and beasts cannot rely on a single dimension such as IQ but instead needs a much broader conception of intelligence.

What Is Intelligence?

Assessing intelligence in machines, humans, and other animals seems to require a definition of intelligence. But definitions are always defeated by counterexamples and circularities, and we learn little from dictionary definitions of intelligence such as "the ability to acquire and apply knowledge" or "the ability to learn or understand." A recent book of essays on intelligence edited by Robert Sternberg includes ten incompatible definitions. The most inane defines intelligence as what IQ tests measure, which is like saying that value is what money measures. There is much more to value than money, and there is much more to intelligence than the linguistic and logical abilities captured by IQ.

Instead of a definition, I follow Greg Murphy and other psychologists in offering three aspects of the concept of intelligence: examples of intelligent people, typical features found in those examples, and explanations provided by the concept. Everyone will agree that intelligent people include Albert Einstein, Marie Curie, Thomas Edison, Jane Austen, Confucius, Ludwig van Beethoven, and Martin Luther King Jr.

We can similarly identify typical features of intelligence without worrying about whether they provide a strict definition that applies to all and only examples of intelligence. Such features include problem solving, learning, understanding, and reasoning. This list is just a start that I expand in chapter 2 to include perceiving, planning, deciding, abstracting, creating, feeling, acting, and communicating.

Like other concepts, intelligence also plays an important explanatory role, telling us what mental capacities enable people, machines, and animals to solve complex problems, learn rapidly, and exhibit creativity. We use the concept of intelligence to explain why some people seem to be faster and more effective at accomplishing tasks, understanding puzzling events, learning, and reasoning. Going in the other direction, psychologists attempt to explain the origins of intelligence in terms of factors such as genetics, social environments, and motivation. Machine intelligence is explained by algorithms that enable computers to solve problems and learn, as shown by outstanding current examples of AI.

Six Smart Machines

The field of machine intelligence began in the 1950s when powerful computers were first developed. Allen Newell, John Clifford Shaw, and Herbert Simon developed a breakthrough AI program called the Logic Theorist in 1956. Initially, progress was slow, as problems that researchers thought would be easy to solve on a computer turned out to be much more challenging. Since 2010, however, the field has taken off with many commercial successes.

Because intelligent computers operate in such different domains, there is no point in trying to rank them based on how smart they are. Instead, here is a nonordered list of what strike me as some of the most impressive accomplishments in intelligent machines.

IBM Watson

In 2011, IBM's Watson computer system competed in the television game show *Jeopardy!* and soundly beat two human champions through its ability to answer a wide range of questions with speed and accuracy. Since then, Watson has become a large enterprise at IBM, with applications that include medicine, business, and law, using techniques that go well beyond the original Watson's ability to answer questions.

Watson's accomplishments match many of the typical features of intelligence. Watson solves problems such as medical diagnosis, and it can learn by generalizing from data, for example, to suggest solutions to business problems such as taxation. Watson's linguistic abilities enable it to process language in both spoken and written forms and to generate arguments. Watson even shows sparks of creativity when Chef Watson generates recipes that are both original and appealing.

DeepMind AlphaZero

By 1997, computer programs could beat human players at chess and checkers. The game of Go is more challenging, because the larger board allows for a much larger number of possible moves. Nevertheless, in 2016 the world's best Go player was defeated by AlphaGo, a program produced by DeepMind, which is now a division of Alphabet, the company that owns Google. AlphaGo even showed flashes of creativity in coming up with moves that surprised human Go experts but turned out to be effective.

Even more impressively, a version called AlphaZero became even more successful without being trained by humans, learning to play just through experience by competing with itself. In four hours of self-training on chess, AlphaZero became capable of beating the world's top AI chess player. DeepMind has also had success at other challenging problems, such as learning to play computer games and to predict how proteins fold.

Self-Driving Cars

People take driving for granted as a simple task accomplished by many sixteen-year-olds, but it requires recognizing objects, planning routes, and learning to drive better. Early attempts at driverless vehicles were unsuccessful because the vehicles could not tell the difference between freeways and exits or between rocks and shadows. The major breakthrough came in 2005, when more than twenty vehicles produced by different teams managed to finish a difficult desert course over a distance of 132 miles (212 km).

Since 2005, the technologies for self-driving cars have steadily improved, enabling them to recognize patterns and make decisions in complex environments. Machine intelligence is already assisting drivers of thousands of Tesla cars, and numerous companies hope to produce completely autonomous vehicles within the next decade. Major competitors include Waymo (a subsidiary of Alphabet), Tesla, General Motors, Ford, BMW, and Uber. I am looking forward to being driven around by a vehicle free of the lapses in attention that are common in human drivers subject to fatigue, distraction, and intoxication.

Alexa and Other Virtual Assistants

I dictated the first draft of this chapter using Dragon NaturallySpeaking, a program produced by the company Nuance. This program enables me to get my ideas down much more efficiently and with less physical strain than writing the old-fashioned way. Speech recognition systems solve the tricky problem of transforming spoken language into text on the computer. I also use Apple's speech recognition system to dictate emails and Amazon's Alexa to play music.

Digital speech recognition began in the 1960s but took fifty years to become effective. Apple's Siri began on the iPhone 4 in 2011. Speech recognition systems are intelligent in that they solve a difficult problem by using machine learning to produce the ability to change speech into text.

Google Translate

To help with my French reading, I often use Google Translate, which works well with words, sentences, and whole documents. Early attempts at machine translation, going back to the 1950s, made too many mistakes to be impressive or useful. Babel Fish, the first web-based translation service, which became available in 1997, produced some laughable results when people translated text into a foreign language and then back again into English. But Google Translate is now widely used across a hundred languages.

Google Translate counts as intelligent because it solves the problem of moving between two languages that is only accomplished by bilingual humans. Polyglots who speak several languages are rare, and no human has ever come close to the hundred languages of Google. Google Translate does not learn as you use it, but much machine learning went into the current version. Smartphone applications are available that provide on-the-fly translations for travelers.

Recommender Systems

If you watch movies on Netflix or order goods from Amazon, then you are familiar with recommender systems that tell you: if you like this item, then you might like that one. Recommender systems now operate in areas that include music, books, restaurants, and online dating. They perform analogical inference, pointing people to items or services similar to ones in which they have already expressed interest.

Recommender systems help to solve problems about what films to watch, what goods to buy, and what dates to find. Lots of these systems operate in the background, for example, when the Netflix recommendation engine is trained to sort movies into categories that it can use to suggest what you might like. Recommender systems also learn when they take into account your ongoing purchases and the ratings that you give to items that you have chosen.

Besides these six intelligent programs, computer scientists today are making many other advances in AI, such as systems for face recognition and playing poker that have made major advances in the last decade. I explore this research in chapter 3, with much more critical analysis.

Six Smart Animals

The rise of machine intelligence undercuts the presumption that something has to be alive to be intelligent. Some panpsychists such as William James claim that consciousness is a property of everything in the universe, which implies that rocks and rivers possess a bit of mind. But a more plausible view of the origins of intelligence looks at the development of life, which began on our planet fewer than four billion years ago. Bacteria and plants are capable of detecting changes in their environments, responding to these changes, and signaling other members of their species. However, I do not consider them as candidates for even rudimentary intelligence, because they lack the capacities for flexible problem solving and learning that I placed among the typical features of intelligence.

From the thousands of species of animals that possess intelligence in this richer sense, I select the following six: bees, octopuses, ravens, dogs, dolphins, and chimpanzees. My list does not imply that these are the only smart animals, or even that they are the smartest. But they are representative of the evolutionary history of life on Earth, covering insects, mollusks, birds, and mammals, including primates.

Bees

Bees have fewer than a million neurons but nevertheless are capable of sophisticated forms of problem solving, learning, and communication. The major problem that bees solve is finding flowers that provide nectar and pollen for manufacturing honey. Their perceptual abilities include vision that can detect the different colors of flowers and an acute sense of smell. They have effective methods of navigation that enable them to fly directly back to their hives when an exploration of local territory encounters promising flowers. Their learning capabilities enable them to be trained by rewards to fly left or right or up or down.

Most impressively, bees are capable of communicating to others where they found nectar and pollen. Information about direction and distance is conveyed by a waggle dance. Bees can observe and interpret the waggle dances of other bees and also respond to the sounds that other bees make by vibrating their bodies. Their abilities to navigate, learn, and communicate offer sufficient evidence that intelligence exists even in insects.

Octopuses

Mollusks such as snails, clams, and oysters are not renowned for their intelligence. But octopuses are the geniuses of the mollusk world, with around 500 million neurons, more than found in mammals such as mice. Unlike mammals and birds, whose neurons are largely concentrated in their brains, many of the octopus's neurons are distributed across its eight arms, giving the arms a high degree of independent activity. Octopus anatomy also differs from mammals in other striking respects, such as the absence of a backbone and the presence of three hearts.

Octopus behavior displays an impressive capacity for problem solving and learning. Although they learn slowly, octopuses have managed to navigate mazes, make visual discriminations, unscrew jars, escape from tanks, recognize different people, and even turn off lights by squirting jets of water at the bulbs.

Ravens

Humans are obviously the smartest mammals, but what are the smartest birds? The prize goes to corvids, the family of birds that includes crows, ravens, and jays. Parrots are another family of smart birds with an impressive ability to learn and use language. The word "birdbrain" is sometimes used to describe a stupid person, but raven brains have more than two billion neurons, as many as dogs and more than cats.

Large brains enable ravens to solve problems such as obtaining and caching food, including moving caches that have been seen by other birds. They can get food by pulling on attached strings and can make both short-term and long-term decisions. Ravens communicate with other ravens by using a combination of voice and body movements.

Dogs

Except for fish, cats and dogs are the most popular pets that people keep for companionship and entertainment. People talk to their pets and ascribe to them complicated mental states such as missing their owners and being annoyed at long absences. The high regard that people have for dogs is indicated by books with titles like *Dogs Are People Too* and *The Genius of Dogs*. The intelligence of dogs is marked by their ability to solve problems on their own and to learn from humans, with border collies and poodles getting the highest rankings.

Dogs can solve problems in the wild by using their perception and memory to hunt for food. But dogs excel in their ability to connect with humans, a capacity selected for by thousands of years of breeding. Dogs can learn by watching humans or other dogs. They readily respond to human gestures such as pointing and can follow voice commands. One border collie learned more than one thousand words functioning as verbal commands. Dogs are attuned to human faces and may even be able to discriminate emotional expressions. Some owners think that dogs can figure out what is going on in their owners' minds.

Dolphins

Dolphins and whales are cetaceans, marine mammals that evolved around fifty million years ago. Their brains are large both in mass and in numbers of neurons, but they often have much larger bodies than humans and therefore surpass them in ratio of body to brain size.

Dolphins and whales can solve complex natural problems involving getting food, finding mates, and coordinating social activities. They have sophisticated abilities to communicate with one another, dolphins using clicks and whales using more melodic songs. Some species of dolphins and whales have been found to recognize themselves in mirrors, suggesting self-awareness.

Chimpanzees

The primates closest to humans in intelligence are chimpanzees with two species: common chimpanzees and bonobos. The two species are anatomically similar but behaviorally different. Whereas the common chimp is patriarchal, the bonobo is matriarchal and relatively egalitarian and nonviolent.

Chimpanzees make tools and use them to solve problems in acquiring food, and they can also deal with other problems that require planning for the future. They are capable of many kinds of learning, including rapid learning through causal understanding of situations. They communicate with one another by vocal sounds and body language, and a few chimps have been taught to use simple forms of human language.

Human Brains

How does human intelligence measure up against these smart machines and animals? You might think that people can do everything that the bots and beasts do, but there are exceptions. Some of the sensory apparatus in nonhumans is superior to what people have. For example, bats use sonar, and some birds and fish can sense electromagnetic signals. Moreover, people lack the laser sensors (lidar) and built-in GPS used by some driverless cars. Computers have been better at doing arithmetic than humans since the 1950s.

Nevertheless, humans are still generally much more intelligent than machines and animals, but what is the basis of this superiority? Religious views claim that our intellectual abilities come by virtue of having a God-given soul, but evidence for the existence of such nonphysical entities is scant. More plausibly, human intelligence derives from our large brains, consisting of approximately eighty-six billion neurons and around the same number of glial cells that provide nourishment and support for the neurons.

A student once asked me: if humans evolved from monkeys, how come there are still monkeys? Biological evolution is a lot more complicated than this question suggests, because humans did not evolve from chimpanzees, our closest relative among the great apes. Rather, humans and chimps had an unknown common ancestor around seven million years ago. Since then, numerous species have evolved, including near-human ones that have gone extinct, such as the Neanderthals.

Why are human brains the smartest on the planet? The answer is not merely their volume or their number of neurons, because whales and elephants surpass humans on both counts. Humans are smarter because more of our neurons are located in the prefrontal cortex, which we use in making more complex decisions than these large mammals can carry out.

Romantics and Killjoys

My aim is to compare and analyze the intelligence of humans, animals, and computers, but there are pitfalls that include a simplistic understanding of human intelligence, which makes it easy for bots and beasts to be just as smart as people. The major impediment to fruitful comparison is exaggeration of the strengths or limitations of bots or beasts.

Machines and animals both have romantic enthusiasts and skeptical killjoys. For machine intelligence, the enthusiasts are people who believe that full artificial intelligence is imminent. Ray Kurzweil writes about the singularity, the tipping point where computer hardware and software will continue to accelerate until machines are much smarter than people. Other enthusiasts about artificial intelligence call themselves transhumanists, eagerly looking forward to the point where human intelligence will be utterly surpassed.

At the other extreme are the skeptics who believe that computer intelligence is fundamentally impossible. One argument rarely expressed is that intelligence is a property of nonphysical souls, a bar that computers could never reach because they are inherently physical. More plausible arguments have come from philosophers who point out substantial differences between earlier generations of computers and the intellectual competence of people, such as having emotions and context-sensitive representations that are about the world.

The question of the relative intelligence of humans and animals also has its enthusiasts and skeptics. Romantics emphasize the commonalities between the minds of humans and other animals. In contrast, killjoys point out the large gap between the intelligent capabilities of humans and what other animals can do. The strongest killjoy arguments claim that humans can do some things that other animals never do, such as using language, employing tools, and having emotions. These arguments have been undermined by increased understanding of the mental capabilities of various animals, but we shall see that substantial differences still exist between the minds of humans and animals. Romantics accuse killjoys of anthropocentrism, elevating humans to a special role at the center of the living world. In return, killjoys accuse romantics of anthropomorphism, assuming that animals are just like people.

My goal is to be neither an enthusiast nor a skeptic about bots and beasts, neither a romantic nor a killjoy. For both machines and animals, I strive to avoid the major pitfalls of overrating or underrating them in comparison with human intelligence. Satisfying this goal requires two procedures: (1) for determining what mental processes operate in nonhumans, and (2) for comparing the nonhuman processes with human ones. These procedures steer a fine course between anthropomorphism and anthropocentrism.

The Attribution Procedure

Figuring out when to attribute mental characteristics to nonhumans is complicated. For animals, we want to be able to determine whether they have mental states, properties, and processes ranging from pain to abstract thinking. For machines, we want to determine what mental characteristics they currently have and also what kinds they are likely to have in the future. We need a sound procedure for determining when it is legitimate to assign a mental characteristic such as thinking and consciousness to an entity such as an animal or machine. The procedure should use the same kind of inference that humans use in attributing mental states to other people, for example, when you infer that someone is in pain or is experiencing an emotion such as joy.

There are no simple deductive arguments available that could rely on general principles such as the following:

If a thing has a specific property, then it has a specific mental characteristic.

There are no known general principles of this kind. Perhaps we could use probability theory to judge if a thing has a mental characteristic, but the required numbers are unknown, such as the probability of a fish being in pain given that it writhes.

We can use observation to identify behaviors, but pain and other mental states are not directly observable in others. The form of reasoning most appropriate for evaluating claims concerning the existence of nonobservable entities is *inference to the best explanation*. For example, you can infer that a friend is in pain because it explains behaviors such as grimacing, moaning, and cursing. In science, Newton inferred that the nonobservable force of gravity exists because it explains much observed evidence, ranging from the motions of the planets to the behavior of objects on Earth that rise and then fall. Moreover, twentieth-century physics brought theories about why gravity works the way it does, including relativity theory and the recently confirmed existence of gravitational waves. When Newton first proposed gravity, science had alternative explanations, such as that the planets are carried around by vortices, but today no competing theories can match gravitational theory for its explanatory breadth and simplicity.

In line with the method of Newton and other scientists, I propose the following Attribution Procedure for determining whether to attribute a mental characteristic to a species of animal or class of machines. As a concrete example, I consider the controversial issue of whether fish feel pain.

The Attribution Procedure proposes the following steps for using explanatory coherence to establish whether or not an animal or machine has a mental state, property, or process.

1. Clearly state a hypothesis concerning a kind of thing and a mental characteristic, for example, that fish feel pain.
2. Compile all the evidence consisting of observations that could be explained by the hypothesis that the thing has the characteristic. For example, injured fish lose problem solving capacity that is restored when they are given painkillers.
3. Also compile evidence that is hard to explain by the hypothesis that the thing has the characteristic, for example, that fish lack brain areas used by mammals to process pain.
4. Assemble alternative hypotheses that have been proposed for explaining the evidence in steps 3 and 4. For example, maybe the injured fish are just distracted and not actually feeling pain.
5. Consider deeper explanations for why the competing hypotheses might be true, usually in the form of mechanisms concerning how the thing has the characteristic. For example, fish have pain receptors in their bodies and have brain areas that receive signals from those receptors.
6. Consider analogies to accepted explanations, for example, the extent to which the explanation of fish behavior by feeling pain is analogous to the explanation of human behavior by feeling pain.
7. Evaluate the overall explanatory coherence of the competing hypotheses with respect to all the evidence, taking into account how much the hypotheses explain and predict, whether they are explained by deeper explanations, the quality of analogies, and the extent to which the hypotheses are simple in the sense of not requiring additional assumptions. For example, we would downgrade the hypothesis that fish appear

to writhe in pain because they are remotely controlled by space aliens, which requires the additional unwarranted hypothesis that there are space aliens and that they have a way of controlling the behavior of fish.

8. Determine whether a hypothesis that the thing has the characteristic is part of the overall best explanation of all the evidence. If so, accept the hypothesis. For example, we can accept the hypothesis that fish are conscious if that is part of the best overall explanatory account, taking into account all the available evidence, all the alternative hypotheses, deepening explanations based on mechanisms, the simplicity of the explanations, and relevant analogies.

This Attribution Procedure works equally well for evaluating mental claims about machines. For example, to evaluate the claim that IBM Watson understands language, we can assemble all the evidence that this claim explains, for example, that Watson answered questions on the game show *Jeopardy!*. But it is also crucial to take into account alternative hypotheses, such as that Watson is only using statistical tricks to fake an understanding of language. Deciding whether or not Watson understands language should be based on the inference to the best explanation of all the available evidence.

The Attribution Procedure is more comprehensive and objective than following simple rules that have been advocated in discussions about the mentality of animals and machines. One of the founders of comparative psychology, C. Lloyd Morgan, proposed the following canon: "In no case may we interpret an action as the outcome of the exercise of a higher psychical faculty, if it can be interpreted as the outcome of the exercise of one which stands lower in the psychological scale." This canon was intended to rule out the attribution of complex mental states such as abstract cognition and emotions if more primitive mental characteristics, such as associative learning, could explain the behavior. However, Morgan moderated his canon: "To this, however, it should be added, lest the range of the principle be misunderstood, that the canon by no means excludes the interpretation of a particular activity in terms of the higher processes if we already have independent evidence of the occurrence of these higher processes in the animal under observation."

Like Occam's razor, Morgan's canon is a principle of simplicity, proposing that we should prefer to explain the behavior of an animal using hypotheses familiar from the behavior of lower animals rather than making additional assumptions about special cognitive abilities. For example, researchers have used the canon to justify explaining the problem solving ability of chimpanzees in terms of trial-and-error learning rather than causality-based insight. But Frans de Waal points out that a gain in simplicity at one location can come at the cost of a loss in simplicity at another, because of the need to provide an evolutionary explanation for the origin of the higher cognition that somehow popped up later in evolutionary history. Accordingly, I see no justification in dogmatically applying Morgan's canon and prefer a more general assessment of the available evidence using the Attribution Procedure.

We should also avoid dogmatism in determining whether current or future machines have mental characteristics such as language understanding and consciousness. Rather than pursuing general arguments, I recommend using the Attribution Procedure to evaluate every case on its own evidential merits, using inference to the best explanation.

It might seem easy to figure out whether a machine has a particular mental capacity, because we can just look at its programming code. But not all machines are driven by easily inspected lines of code. For example, computers trained by deep learning have enormous neural networks where the knowledge is distributed among thousands of connection weights. Moreover, some of the mental characteristics can be hard to identify even in explicit code because they will require interactions among different modules in the overall computer program. For example, language understanding is not the simple result of an identifiable subroutine in a long program but requires interactions among different subroutines. Additionally, for machines to have versions of phenomena such as pain, emotion, and consciousness, they may need interactions between their software and hardware, including sensory input. Understanding a complex mechanism is not always straightforward.

Two kinds of errors can arise in ascribing mental characteristics to animals or machines. A false-positive error consists of judging that an entity has a mental characteristic that it lacks, for example, saying that rocks have emotions. A false-negative error consists of judging that a thing lacks a mental characteristic that it actually has, for example, saying that people do not have emotions. Whether it is worse to make false-positive or false-negative errors depends on the consequences of the two kinds of errors. False-positive errors are bad if they prevent people from acting in ways that are otherwise in their interests, for example, in refusing to use rocks as paving stones because their feelings might be hurt. False-negative errors are bad if they lead to inappropriate treatment of things, for example, slaughtering animals brutally on the misconception that they feel no pain or emotions.

The burden of proof in the Attribution Procedure depends on the consequences of making the different kinds of errors. In legal trials in the common law system used in Great Britain, the United States, and other countries, the prosecution is required to show that an accused is guilty beyond a reasonable doubt. The courts use a presumption of innocence because of the ethical judgment that it is worse to convict an innocent person than to acquit a guilty person. However, in civil trials such as lawsuits, the burden of proof is merely the preponderance of evidence, and the court finds for the plaintiff or defendant based on which side the evidence supports. The plaintiff does not need to show that the defendant is at fault beyond a reasonable doubt. So in civil cases, false positives and false negatives are treated as equally bad.

The burden of proof in ascribing mental characteristics to bots and beasts is complicated. Advocates of animal rights place a high value on avoiding mistreatment of animals who are capable of suffering, so they feel strongly that false negatives are worse than false positives. They might say, for example, that we would want to establish beyond a reasonable doubt that fish do not have consciousness so that we can avoid the dire consequences of making them feel pain. On the other side, an avid meat eater might argue that false positives can be bad too, for example, if people are unnecessarily deprived of nutrition and enjoyment because humanlike pain and emotions are assigned to animals without sufficient evidence. These deliberations are not governed by overriding values akin to not convicting the innocent, so preponderance of evidence is the appropriate standard, rather than reasonable doubt. Careful application of the Attribution Procedure considers a full range of hypotheses and evidence but does not guarantee conclusions that are free of errors, although it should make them less common.

Using the Attribution Procedure is also superior to applying general principles that purport to settle issues about attribution of mental characteristics. At one extreme, we have the philosophical position of panpsychism, which assumes that everything in the universe has some mental characteristics despite the lack of evidence of mind in atoms and dirt. At the other extreme, we have the scientific and philosophical approach called behaviorism, which says that we should never attribute mental characteristics to anything. Behaviorists claim that postulating the existence of nonobservable entities such as mental states and processes is unscientific. This claim has been undermined within science because of the failure of behaviorist psychology and linguistics to predict and explain the behavior of humans and other organisms. The claim is also undermined by philosophy of science, which has noticed that the most well-established sciences explain observations by hypotheses concerning nonobservable entities and processes such as electrons, forces, atomic bonds, gravity, genes, and viruses.

The Comparison Procedure

I am concerned not just with whether bots and beast have mental characteristics but also with how their minds compare systematically with those of humans. I use the following Comparison Procedure.

1. Develop a rich account of human intelligence by specifying features that describe it and mechanisms that explain how humans have the features.
2. Assess the extent to which examples of machine intelligence display these features and mechanisms.
3. Assess the extent to which examples of animal intelligence display these features and mechanisms.

This procedure provides a multidimensional evaluation of the comparative intelligence of numerous animals and machines that goes far beyond what IQ tests could offer. The aim of the comparison is to learn about the current state and ethical significance of intelligence across all these entities.

The first step in assessing the intelligence of machines and animals is to construct a theory of human intelligence. Chapter 2 examines the accomplishments of some of the most impressive humans, considering problem solving and learning in multiple domains such as science, technology, and the arts. A theory of intelligence should go beyond listing the observable features of different kinds of intelligence to state the fundamental mental mechanisms that generate problem solving and learning in all these domains. Human intelligence goes beyond linguistic reasoning to encompass sensory imagery, emotions, and consciousness.

This account of twelve features and eight mechanisms yields a twenty-item checklist for assessing intelligence in bots and beasts. For each machine or animal, the Comparative Procedure generates a report card that summarizes how well it stacks up against human intelligence. The intelligence checklist provides a tool for critically examining the current accomplishments in, and prospects for, machine and animal intelligence.

Chapter 3 applies the same checklist to six smart machines to scrutinize what they accomplish and how they accomplish it. For each machine, I evaluate its performance in its particular domain and provide an informal description of the computations that produce its performance. No one of these machines comes even close to human intelligence, but taken together, they show how machine intelligence today measures up against human intelligence.

Chapter 4 performs the same evaluation for six smart animals. The intelligence checklist and the resulting report card reveal the strengths and limitations of animal thinking. The point of these comparisons is not to declare winners and losers but rather to illuminate how intelligence works across humans, animals, and machines. Chapter 5 provides a more general assessment of the mental advantages that people have compared to other animals, which also illuminates why machine intelligence continues to fall well short of human performance. Chapter 6 applies the Attribution Procedure to controversial issues about animal minds concerning bacterial consciousness, fish pain, dog jealousy, and ape analogy. It also explains the limitations of the Turing test for evaluating machine minds.

Chapter 7 examines the ethical issues that arise with appreciation of the intelligence of current and future bots and beasts. It uses a common ethical framework based on needs to evaluate threats to the existence and well-being of humans, animals, and possibly machines. Topics include the robot apocalypse, animal extinction, vegetarianism, pets, automation, privacy, social control, and killer robots. The principle "need, not greed" provides guidance on the development of policies for social progress. Chapter 8 deepens the ethics of artificial intelligence by using principles from medical ethics to organize and assess the principles and values relevant to current and future developments in AI.

In the title of a book, Frans de Waal asks, "Are we smart enough to know how smart animals are?" My answer is yes, if we use the Attribution and Comparison Procedures to approach the question thoroughly. The answer is also yes for whether we are smart enough to know how smart machines currently are, but undecided for whether smart machines will surpass humans in the future. My main conclusions are that a full appreciation of human minds shows that bots have a long way to go before they threaten human intelligence, and beasts similarly fall well short of human capabilities. People are smarter than you think.

I have yet to answer specific questions about cat jealousy, fish pain, Alexa's happiness, and self-driving cars. Later chapters use the Attribution Procedure to suggest that Alexa is neither happy nor unhappy, cars can drive themselves, Pixie probably is jealous, and fish may feel pain.

2 Prodigious People

When my sons and I visited the Rodin Museum in Paris in 2004, Adam and Dan spontaneously imitated *The Thinker* (fig. 2.1). This statute has powerfully represented the human intellect since 1904, displaying the force of concentrated thought. Rodin needed intelligence to create such a striking image of intelligence. What makes people smart enough to compare intelligence across humans, animals, and machines?

In this chapter, I expand the initial characterization of intelligence in chapter 1 by adding a larger set of examples of intelligent people across many domains of human activity in areas such as science, technology, the arts, and social organizations. Even mundane tasks such as shopping for groceries require some intelligence: you would not send a pet, robot, or young child to the supermarket to buy the ingredients for a meal. Grocery shopping illustrates the twelve typical features of intelligence listed in chapter 1: perceiving, problem solving, planning, deciding, understanding, learning, abstracting, creating, reasoning, feeling, acting, and communicating.

How are people capable of carrying out these functions? An explanation of intelligence should be more than just a story about kinds of intelligence; it should describe what goes on in your mind and brain that enables you to perform all the mental procedures that constitute intelligence. Today it seems obvious that human thinking depends on brains, but this discovery was made by the ancient Greeks, Alcmaeon and Hippocrates, around 2,500 years ago. Contemporary cognitive science explains how minds work by using both mental mechanisms and neural mechanisms. Mechanisms are combinations of connected parts whose interactions produce regular changes, for example, when the connections and interactions of a car's engine, transmission, and wheels move it along a road.

Figure 2.1
Adam and Dan Thagard in Paris in 2004. Photo by Paul Thagard.

Laying out the features and mechanisms of human intelligence provides a checklist for comparing people with machines and animals. We can inquire whether animals and machines can carry out the functions that are typical features of human intelligence, and whether they depend on the mechanisms that enable people to have those features.

Examples and Domains of Human Intelligence

People use their intelligence in many daily activities. We have to navigate the world to find food, and we have to avoid dangers such as speeding cars, although today we rarely need to avoid predators who want to make us into their food. Life is social, so people have to figure out how to get along with others sufficiently well that they cooperate with us. The physical act of sex does not require much intelligence, but finding a mate can be a complicated process, as in internet dating. More advanced intelligence shines in four domains: social innovation, technological invention, artistic imagination, and scientific discovery.

Social Innovation

One of the reasons why brains became larger over the millions of years of primate and human evolution is the complexity of social interactions. Many people are exemplary in their ability to acquire knowledge and skills for solving problems and learning in the social domain. The social domain has led to many creative innovations such as cities, churches, governments, and universities.

Here are some examples that I find impressive. Mahatma Gandhi used nonviolent means to unify the diverse peoples of India and overthrow their British rulers. Angela Merkel displayed substantial social and organizational skills in running Germany for more than a decade and carrying out difficult decisions such as admitting more than a million refugees in 2015. Franklin Delano Roosevelt made substantial changes to American society to deal with the Great Depression of the 1930s. Florence Nightingale required substantial organizational and social abilities to bring about fundamental changes in the practice and recognition of the nursing profession in the nineteenth century. Generate your own list of people who have displayed substantial intelligence in working toward social change.

Technological Invention

The American humorist Dave Barry pronounced: "Without question, the greatest invention in the history of mankind is beer. Oh, I grant you that the wheel was also a fine invention, but the wheel does not go nearly as well with pizza." No one knows who produced early technological innovations such as the stone ax, the wheel, beer, and pizza. But human intelligence is exemplified by hundreds of inventors who produced valuable products from the printing press to the Google search engine. Early humans used tools such as sharp stones, and hundreds of thousands of years of development have brought new tools that are both applications of intelligence and enhancers of intelligence. Inventions such as agriculture, the wheel, paper, printing presses, and computers mark high points in human intelligence.

Here are some exemplars of human intelligence in the realm of technology. Thomas Edison contributed to the development of hundreds of products, including the electric light bulb, the phonograph, and the movie camera. Grace Hopper was an early computer scientist who invented the first computer compiler and one of the earliest high-level computer languages. Soichiro Honda helped develop innovative motorcycles and cars. George Washington Carver invented many peanut products, although not peanut butter. All these highly intelligent inventions required substantial amounts of problem solving, learning, creativity, and other features of intelligence.

Artistic Imagination

The arts are not always associated with intelligence because of the mistaken limitation of intelligence to linguistic and logical abilities. But knowledge, skill, and learning are hardly restricted to language and logic. Figuring out how to produce artistic works that are beautiful or emotionally evocative in other ways requires substantial intelligence, connected with perception, imagery, and other mental faculties.

By forty thousand years ago, humans were also displaying intelligence in artistic activities such as cave paintings and musical instruments. Today we have many other fields of artistic imagination, including literature, sculpture, dance, photography, and architecture. Merely copying what others have done requires only modest intelligence, but producing works that are new and emotionally powerful requires features of intelligence like problem solving, learning, and creativity.

Hundreds of intelligent artists could be marked as exemplary, but here are a few of my personal favorites. Beethoven produced musical masterpieces that are still regarded as creative and accomplished. I view Georgia O'Keeffe as one of the great painters of the twentieth century because of her beautiful and surprising depictions of landscapes and flowers. Ahmad Lahori was the chief architect of one of the world's most beautiful buildings, the Taj Mahal. Ray Charles was viewed as a musical genius because of the innovative ways in which he combined musical styles, including gospel, blues, country, and pop. Such creative artists exemplify intelligence in learning to get better at solving complex aesthetic problems.

Scientific Discovery

The fourth major domain of human intelligence is science, which developed much more recently than social interactions, technology, and art. Observational science as a systematic enterprise began with the Babylonians around 3,000 years ago, and theoretical science independent of theology began with the ancient Greeks around 2,500 years ago. Since then, science has been a spectacular manifestation of human intelligence, with steadily increasing knowledge in burgeoning fields that include physics, chemistry, biology, and the social sciences.

Scientists generally have high IQs, but their intelligence is displayed in many ways besides doing well on tests. Being a successful scientist requires a high capability for solving complex problems and rapid learning. One can quickly generate hundreds of exemplars of scientists as intelligent thinkers. I mentioned some of my favorites in chapter 1, and another example is Charles Darwin, who created the theory of evolution by natural selection that still informs all of biology. Besides Marie Curie, who was awarded not one but two Nobel Prizes, another Nobel Prize winner was Tu Youyou, a Chinese woman who discovered the drug artemisinin, which works as a treatment for malaria. Neil deGrasse Tyson is an astrophysicist and successful science communicator.

For understanding a concept, listing exemplars starts to pin down its meaning by giving standard instances that most people recognize, even if definitions are hard to specify. But a richer understanding of a concept requires identifying its typical features that apply across many domains.

Features of Intelligence

I remember my mother saying that she routinely sent me to the grocery store for bread when I was two years old. The grocery store was just at the end of the street, and I was precocious, but I still doubt whether the story could have been true. Going to the store requires solving a series of problems that are beyond the range of toddlers, including navigating up the street, communicating to the store clerk what to buy, and handling payment. Grocery shopping is hardly the pinnacle of intelligence, but because of its familiarity, I use it to illustrate the typical features of intelligence, complemented by their occurrence in art and science.

Perceiving

Sensing the environment does not require intelligence, as even bacteria and trees can detect changes in chemical and light sources. Perceiving is more complicated in requiring inferences about what is being sensed, for example, when you see that an object is a bird rather than a branch. When you go to the grocery store, many perceptions are required, especially using the senses of seeing and smelling. You look at the vegetables to see which look fresh, and smell the fruit to see whether they are local or have been in storage for months. Touch is important for detecting whether tomatoes, peaches, and other ingredients are ripe. Sound can also matter when you shake a coconut to detect how much milk it contains. You are not supposed to taste the items in the grocery store, but sometimes samples are available, and I have seen people surreptitiously taste cherries to see if they are sweet. Making intelligent decisions at the grocery store depends on having accurate perceptions of many foods.

Like the other typical features, perceiving is neither necessary nor sufficient for intelligence because some activities, like proving theorems in mathematics, can be done without it. But dealing with other people, art, technology, and science usually requires perceptions to recognize situations for problem solving and learning. Galileo needed vision to identify the rings of Saturn, and Mozart needed hearing to compose his operas. Seeing, hearing, and the other human senses are not just the passive reception of signals but require knowledge-based inference to interpret the world. So perception deserves recognition as one of the typical features of intelligence.

Problem Solving

People solve problems when they have goals and try to figure out how to accomplish them. In the grocery store, your main goals are finding good food to eat and avoiding excessive spending. These goals may be incompatible if your favorite foods include expensive items such as beef tenderloin and halibut. Other constraints on your grocery shopping include the time available and the need to manage children who accompany you. I once went shopping with my son Adam when he was around five years old. His inquisitive nature led him to see what happens when he lifted the handle on the coffee bean dispenser, and beans spilled noisily all over the floor.

The grocery store gives you many options for accomplishing your goals of buying food. Sometimes, if you have a specific recipe in mind, your goal is to find a set of foods from the recipe. The search can be routine if you have shopped in the same grocery store, but otherwise you have to search the aisles to find the items you want. A successful shop is when you leave the store with all the items that you wanted and have not exceeded your budget. People can flexibly adapt to solving new kinds of problems, such as keeping distance from other shoppers during the COVID-19 pandemic.

Artists and scientists have more difficult problems to solve. A painter may have to generate a good number of attractive paintings before a show deadline, subject to constraints such as maintaining aesthetic standards and venturing in some new directions. A biologist may need to figure out how to design a new experiment to test an important theory but also do so within the constraints of budget and personnel. In these cases, intelligence serves to accomplish goals while satisfying constraints. Three kinds of problem solving are sufficiently important to count as features in themselves: planning, deciding, and understanding.

Planning

Planning is a kind of problem solving that involves working out a series of actions that will accomplish goals over an extended period of time. When I was the single parent of two hungry teenagers, I worked out a shopping list system that people react to with either admiration or horror. I got a clipboard, stuck it to my refrigerator with magnetic tape, and attached to the clipboard a shopping list, printed in Microsoft Word, organized according to the layout of my favorite grocery store. Then all I had to do for each shopping trip was to circle the items that I wanted to buy, with the list capturing exactly the order of appearance of the items in the grocery store, starting with the deli items and ending with the frozen foods. Planning a trip to the grocery store thus became simply a matter of circling the items that indicate the series of actions to take in the store. COVID-19 made grocery planning trickier because of the goal to shop online.

Without planning, grocery shopping would be a random process of wandering through the store and picking up whatever items look good to you. Such problem solving is inefficient, wasting time and making it unlikely that you return home with the items you want. Planning similarly adds enormously to the effectiveness of many human activities involving social organization, science, technology, and even the arts. For example, urban planners lay out rules and actions concerning how cities can deal with issues about housing, transportation, and communication. Scientists plan research programs consisting of series of experiments and theoretical interpretations. Engineers plan future designs, for example, what smartphones to develop over the next few years. Artists tend to be more spontaneous but can still develop plans about what books, plays, paintings, sculptures, song albums, or other projects to pursue.

Humans have large enough brains that we do not have to solve problems opportunistically only in the moment but instead can imagine how events might unfold over time. Imagination enables us to come up with better plans that produce better solutions to problems.

Deciding

Planning sometimes narrows down to a choice between specific actions, as when grocery shopping requires you to make decisions such as whether to buy broccoli or cauliflower or which kind of cheese to pick up. Deciding is usually an intelligent activity, because it requires you to make a complicated inference about which potential action best satisfies your goal. Before going grocery shopping, you already had to decide which grocery store to shop at, based on constraints such as distance and quality of food.

Political leaders make decisions such as how much money to spend on social programs. Scientists and engineers sometimes choose between alternative research programs based on which seems more intellectually promising and fundable. Artists make decisions daily about how to modify a work in progress to improve it. All these examples require evaluating the merits of a small number of options to choose from.

Deciding is mentally challenging for two reasons. First, you have to keep in mind different possible actions to evaluate which offers the best way of accomplishing your goals. Second, accomplishing your goals may require complicated trade-offs, for example, balancing quality of food versus cost of food. Many animals and machines are capable of problem solving, but planning is much rarer because of the need to imagine scenarios over time. Decision making is also rare because of the mental demands of systematically comparing two or more options.

People are not always good at deciding. We are impulsive in doing the first thing that comes to mind without balancing our goals, for example, when I grab a bag of chocolate-covered almonds at the grocery store while forgetting the constraint of healthy nutrition. But when people consider multiple options subject to careful evaluation, deciding well is a hallmark of intelligence.

Understanding

Another important kind of problem solving involves finding explanations that provide understanding. Often I go grocery shopping on automatic pilot, just picking up what I need at the moment. But sometimes puzzling events occur that make me look for explanations. Why has my supermarket replaced some of the regular checkout stands with automatic ones where people have to scan their own purchases? What happened to the fresh lobster tank that used to be in the fish section? Why are only some of the checkout stands available for buying beer? How and when does the supermarket staff restock the shelves?

The need for understanding is sometimes prompted linguistically. Figure 2.2 depicts an ambiguous sign I saw in 2019 at the University of Bristol. The explanation of the sign is not that there is something called a "philosophy disability" but rather that the Philosophy Department and the Department of Disability Services are in the same lane. People often want not only to know what is happening but also to understand why it is happening. This desire is especially strong in science and technology, but it also operates in social organizations where people want to know why individuals and corporations are behaving oddly. Explanation and understanding are

Figure 2.2
Sign at the University of Bristol. Photo by Paul Thagard.

also relevant to the arts, for example, when literary scholars wonder why a writer used a particular style.

Explanations can take different forms, such as stories that are commonly a part of human culture. But explanations in science and technology usually go deeper, relying on mathematical principles or descriptions of causal mechanisms that indicate how systems work through the interactions of their parts. Then understanding requires identifying causes of puzzling events, for example, explaining why people with influenza are sick by identifying the virus that causes their symptoms.

Learning

An intelligent problem solver should not just do the same thing over and over again but should learn to solve problems better through experience. Many kinds of learning are relevant to becoming a better shopper. You can learn from experience to make generalizations, for example, that bananas with brown spots are overripe. Some of your learning is based on rewards and punishments, for example, when you find a food that tastes wonderful, such as really fresh strawberries, or when you find a food that tastes repulsive, such as a particularly stinky cheese.

Other kinds of learning relevant to grocery shopping include acquiring new concepts, for example, when you learn that a plumcot is a hybrid of plums and apricots. COVID-19 introduced new concepts such as *social distancing* and *quarantine haircut.* Your learning can sometimes also involve forming new hypotheses, for example, the reason that the store's produce is limited on Mondays is that the staff usually restock the shelves on Friday for the weekend. Another kind of learning that is valuable for improving your grocery problem solving is figuring out how to navigate the store, for example, arranging to pick up frozen foods at the end of your shop so that they do not melt before you get home.

Learning contributes to better problem solving in all domains. Social organizers should learn how to better meet the needs of their clients. Scientists learn better ways of designing experiments and interpreting their results by using statistical techniques and causal reasoning. Engineers and other technologists learn causal patterns that guide the production of better machines. Artists learn new techniques such as better ways of producing paints and brushstrokes.

Here are some of the kinds of learning that contribute to human intelligence:

1. Associative learning: bananas are yellow.
2. Reinforcement learning: peeling bananas makes them taste better.
3. Analogical learning: plantains are like bananas and so can be peeled.
4. Causal learning: bananas become brown because they are overripe.
5. Hidden-cause learning: bananas have a strong scent and taste because they contain isoamyl acetate produced by the gene AFT1.
6. Learning by imitation: one child sees another child peel a banana.
7. Learning by being taught: parents show their children how to peel bananas.

Recognizing hidden causes requires the process of inference to the best explanation that goes beyond what is perceivable.

My comparison of humans with machines and animals will grade them on the extent to which they are capable of all these kinds of learning. Associative learning and reinforcement learning require many repetitions, but people can learn from analogies, causes, imitation, and teaching based on a small number of cases, sometimes only one. For example, if the store manager tells you that peanuts are in the produce section, you may not need to be told again.

Abstracting

Advanced learning usually requires forming abstractions. Shopping at a grocery store can mostly be done by perception, where you see something such as a banana or cauliflower and put it in your basket. However, humans are not limited to what we can sense and perceive, because we can abstract from what is immediately before us and form more general categories. You can learn concepts such as *apple* and *orange* just by seeing apples and oranges, but the more general and abstract concept *fruit* requires a higher inference to identify what is common between apples and oranges, such as having seeds that produce the trees on which the fruits grow. Other abstract concepts that are useful during grocery shopping include *vegetable, meat, cold cereal, hot cereal, soda*, and so on. Even more abstract distinctions include produce, dry goods, beverage, and stock-keeping unit (SKU).

You also go beyond sensory experience when you show a concern for the safety of foods. You want to avoid meat, seafood, and other foods that have been contaminated by bacteria and viruses, even though you cannot see germs. Perhaps you spend extra money on organic food because you believe that pesticides are dangerous for your health. Moreover, when you pay for your groceries, you do so with money, an abstract concept that depends on social conventions about how concrete objects such as food and clothing can be exchanged for something that nobody observes. Money in the form of paper currency and coins is already an abstraction, but the money you used to pay for your groceries by using your credit card or debit card is even more removed from the senses.

Scientists revel in abstract concepts such as *atom, force, relativity, cell, gene,* and *natural selection,* all of which go beyond sensory experience to explain important observations. Such abstraction existed long before science developed, as religious concepts found in many cultures such as *spirit, god,* and *angel* postulate causes that go beyond perception. Artists also surpass the senses with abstract concepts like *form* and *beauty.*

Here are the kinds of abstracting performed in human learning:

1. Forming concepts that capture common perceived properties of objects, e.g., *apple.*
2. Forming concepts that capture perceived relational properties of objects, e.g., *smaller–larger, same–different, above–below,* and *edible–inedible.*
3. Forming concepts that capture hidden causes of objects, e.g., *molecule, gene, belief, god.*

We will see that animals and current machines fall short on this third kind of abstraction.

Creating

You do not need to be creative to be an effective grocery shopper, and in general creativity is not a necessary condition of intelligence. Creativity always requires learning, but you can learn without creativity. Nevertheless, many of the activities that are considered to be the most intelligent in scientific discovery, technological invention, artistic imagination, and social innovation are rich in creativity.

A product such as an invention is creative if it is new, valuable, and surprising. Occasionally creativity can enter into grocery shopping when you buy products with a novel plan to put them together in a way that people will like. For example, I sometimes go shopping without a definite list but with a plan to pick up whatever produce seems freshest and most attractive. I do not need a recipe for fruit salad because I pick out the fruit that looks best, which may lead to novel combinations. One new combination that I came up with uses ripe mango, peach, and blueberries, drizzled with melted dark chocolate. What I produced was novel for me (and apparently for the internet), and it was valuable because my dinner guests all liked it. They also found it surprising because of the unusual combination.

So my grocery shopping turned out to be a creative enterprise. People in general are not creative every day of their lives, but scientists are creative when they come up with new experimental designs and generate novel hypotheses that explain experimental results such as relativity theory. Engineers are creative when they invent valuable new devices such as the air conditioner. Social innovation is creative in producing new institutions such as universities and new programs such as universal health care. Artistic creativity is evident in particular products such as painting and books and also in new methods such as cubism and free verse.

We can grade creativity according to the extent to which the resulting things, concepts, hypotheses, or methods are new, surprising, and valuable. A genius is someone who generates many highly creative products.

Reasoning

All thinking requires inference, which I understand as the transformation of mental representations such as concepts and beliefs into new representations. Reasoning is sometimes taken as equivalent to inference, but I view them as different. Inference is carried out by the brain, with billions of neurons operating in parallel, and is often unconscious. It can operate on nonverbal representations such as visual images, as when you find your lost umbrella by visualizing where you might have left it. In contrast, reasoning is verbal, conscious, and social when we use language to give reasons to other people about what they should believe or do. Inference is too ubiquitous to be considered a mark of intelligence, but good reasoning is one way that people impress others by their intelligence.

Philosophers classify reasoning as deductive, inductive, or abductive (explanatory). Deduction goes from premises to conclusions without introducing uncertainty, as in the pattern called modus ponens: If that is an apple, then it has seeds; that is an apple; therefore it has seeds. Induction introduces uncertainty, as in the generalization "Every apple I have seen has had seeds, so all apples have seeds." This conclusion fails because the web tells me that someone has come up with a seedless apple.

Finally, abductive inference generates hypotheses to explain puzzling events, for example, concluding that a genetic mutation produced a seed-free apple. Physicians reason abductively when they reach diagnoses about what diseases cause their patients' symptoms, as do mechanics when they figure out what broken part is responsible for your car not working. Abductive reasoning requires explanations that usually depend on an understanding of causality.

All three kinds of reasoning are important for scientists, who deduce the experimental predictions of their theories, induce experimental phenomena from observed results, and abduce hypothetical causes that explain the results. Similarly, engineers use all three kinds of reasoning in solving technological problems and devising new solutions. Deduction is probably rare in social planning, but inductive learning from experience and abductive generation and acceptance of causal hypotheses are important for designing programs such as health-care systems. Artists perform at least some inductive and deductive reasoning in verbally describing their nonverbal productions.

Feeling

Stereotypically, feelings run counter to intelligence, as when people are asked whether they are being rational or emotional. Plato and many later thinkers thought that a crucial part of intelligence was keeping the emotions in check. But this common view has been exploded by numerous findings in psychology and neuroscience that identify how intelligent problem solving and emotional feeling are entwined.

Several kinds of conscious experiences contribute to effective grocery shopping. Perceiving is not just a matter of coldly recognizing that the peaches are ripe, but can also involve the experience of noticing how delightful the peach smells. Perceptions are not just inferences but also experiences that combine sensations with emotional evaluations to produce feelings.

Emotions also contribute to grocery stopping when they affect decisions. You may reach for a piece of cheese but recoil when you realize how expensive it is. Saving money feels good because it helps you to accomplish your goals of spending responsibly, whereas being extravagant may make you feel guilty or even ashamed. More positively, feeling good about buying something can lead you to actually do it in ways that are consistent with your goals of getting food that is healthy and pleasurable for you and the people you are feeding. You may be encouraged to buy something new by emotions such as curiosity and boredom with past habits. So intelligent grocery shopping is not just a matter of thinking and reasoning but can also be productively caught up with emotional feelings. When you shop, you apply values, such as eating well and saving money, that are not just preferences; they are emotional goals that explain why you prefer one purchase over another.

Emotional intelligence has become a major topic in organizations because an individual's success depends on more than raw thinking power. Many social problems require the ability to understand and control one's own emotions and those of others through self-awareness, self-regulation, empathy, and motivation, all of which employ feelings. Scientists today rarely work alone, and they require intelligent feelings to deal with their collaborators and to motivate the intense effort that research requires. Artists use feelings such as beauty to guide their own work and to influence interactions with their patrons and audiences. So feeling often serves as a feature of intelligence rather than as an impediment.

Communicating

At my grocery store, I sometimes see people shopping in pairs, stopping to discuss what items to buy. Occasionally people stop in the aisle to call home and check on what needs to be purchased. Communication also takes place during grocery shopping when people ask staff members for advice, for example, concerning whether fish is fresh or whether a processed food contains peanuts. During COVID-19, stores have painted arrows on the floor to communicate with shoppers about how to stay safe.

Communication is a minor part of grocery shopping but is integral to many other intelligent activities. Most scientific work today is done by teams, and almost all scientific publications are coauthored. Running experiments, building instruments, interpreting data, and writing up papers are all processes that require communication. Similarly, social organization and technology production are rarely individual activities but require people to have lengthy and productive communications. Artistic imagination is more likely to be an individual activity, but artists such as painters and musicians often benefit from communicative contributions from other people. For example, Pablo Picasso and Georges Braque developed cubism together while sharing a studio for seven years.

Group intelligence is more than just the sum of the intelligence of the people in the group, because the interactions of its members can lead to accomplishments that the members could not produce individually. So part of being intelligent is being able to communicate with other people and to combine expertise with joint results that sometimes qualify as creative.

Acting

Once you have performed perceiving, problem solving, and learning, it may seem easy to carry out the actions that your deliberations have supported. Some actions may be automatic through reflex, instinct, or immediate precedent. But actions can also depend on intelligence, for example, when you have to figure out how to get your arm to reach a box of cereal on the top grocery shelf, or when moving your shopping cart requires squeezing through a narrow passage between two other shoppers.

Grocery shopping is not as complicated as ballet dancing or basketball, but it still requires coordination of your feet and hands to get what you want. Similarly, painters and sculptors need to be able to figure out how to use their bodies to accomplish their aesthetic goals. Intelligence in action is also an important contributor to the use of instruments by scientists in laboratories and by surgeons in operating rooms, and in the use of tools by mechanics to fix cars. Hence physically acting is an important feature of intelligence that we can take into account in comparisons with robots and animals.

I have proposed twelve typical features of intelligence: perceiving, problem solving, planning, deciding, understanding, reasoning, learning, abstracting, creating, feeling, communicating, and acting. This list is subject to amendment but provides a provisional checklist for the comparative intelligence of machines and animals. It also provides a set of explanatory targets for an adequate theory of intelligence that should be able to specify how humans have those features. Current theories of intelligence fail to provide such specifications, but mental mechanisms can explain the origins of all these features.

Theories of Intelligence

What is a scientific theory? In physics and economics, theories are usually sets of mathematical equations that can be used to generate predictions. In history and some social sciences, theories are often just verbal narratives about how things come about. But in biology and psychology, the best available theories are descriptions of mechanisms. In biology, how bodies work is explained by cell theory, where the parts are cells that are chemically connected to form tissues and organs that make bodies function. Neuroscience studies mechanisms where the parts are neurons, the connections between neurons are synapses, and the interactions consist of neurons exciting and inhibiting one another to produce patterns of firing.

In cognitive psychology, our understanding of the neural underpinnings of thought is rapidly increasing, but it is also useful to investigate mental mechanisms without diving all the way down to neurons. Then the parts are mental representations such as images, concepts, and rules, and the interactions are computational processes that operate on representations to produce inferences. For example, you infer how many ears a cat has by deducing that cats are mammals and all mammals have two ears. Or you can recall mental pictures of your favorite cats and count the number of ears on their heads.

Most intelligence researchers missed the mechanistic turn that psychology took with the development of cognitive psychology in the 1960s and cognitive neuroscience in the 1980s. They sometimes mark speed of processing and working memory capacity as contributing to intelligence but neglect the neural or mental mechanisms that are responsible for these components. Work on IQ is largely empirical and statistical, although some researchers propose that people have a general factor called *g* that is responsible for people's abilities to do well on various IQ tests and school tasks. But *g* is rarely spelled out in terms of mental or neural mechanisms.

Robert Sternberg and Howard Gardner are famous for their accounts of intelligence broader than IQ but offer taxonomies rather than explanatory theories. Sternberg's "triarchic theory" divides intelligence into analytic (what IQ measures), creative, and practical but is vague about the processes by which these divisions work. Similarly, Gardner's "multiple intelligences" go beyond the logical and linguistic components of IQ to include mathematical, visual-spatial, bodily-kinesthetic, musical, interpersonal, and intrapersonal intelligence. These extensions are useful because human brains have different ways of being intelligent, in the sense of solving problems and learning. But Sternberg and Gardner provide no mechanistic account of how these ways operate.

In contrast, theories in cognitive science describe how problem solving and learning operate using mental and neural mechanisms. These theories are usually framed as "cognitive architectures" rather than as theories of intelligence, which historically is associated with IQ. I will not attempt to review the range of proposed cognitive architectures but instead sketch the theory that I think is most powerful and indicate how it provides explanations for the twelve features of intelligence.

The difference between features and mechanisms of intelligence is like the difference in medicine between symptoms and diseases. Features and symptoms are descriptions that cry out to be explained by mechanisms. For example, if you have a skin rash and intense pain, your doctor might explain these symptoms by the hypothesis that you have the disease of shingles, resulting from the underlying mechanism of reactivation of the herpes zoster virus that originated as chicken pox. Just as diseases explain symptoms, so mental and neural mechanisms should be able to explain the features of intelligence.

Mental Mechanisms

Vague constructs such as *g* for general intelligence and rough taxonomies such as multiple intelligences do not explain much. Fortunately, cognitive science has already identified a plausible set of mechanisms with the appropriate range of application. My 2019 book *Brain-Mind* reviews them in detail, but here I just outline how mind and intelligence can be explained by mental mechanisms of images, concepts, rules, analogies, emotions, consciousness, intentional actions, and language.

For the eight mental mechanisms that follow, I characterize how they contribute to grocery shopping and also to intelligence test results. Then individual differences in functioning of these mechanisms explain differences in intelligence and IQ test performance.

Images

Imagine your favorite grocery store and try to answer the following question: are the apples in the same row as the tomatoes? When I do this exercise, I can see that the apples are two rows away from the location of the tomatoes. Some people may perform this task by verbal reasoning, for example, by thinking that the apples are in row 2, the tomatoes are in row 4, $4-2=2$, so the apple row is two rows away from the tomato row. But for me and most other people, the mental picture of the rows answers the question without any verbal reasoning.

Imagery also operates with other senses besides vision. If you are thinking about putting apples and bananas together in a fruit salad, you can imagine how apples taste, how bananas taste, and how they would taste together. If you hear your name over the store sound system, you can repeat the announcement in your head to try to determine whether your name was actually being called. You can smell a cantaloupe and try to compare that smell to your memories of previous ripe and unripe cantaloupes. Touch imagery can operate if you are thinking of buying a towel and feel it to see if it is as soft as your favorite towels. Mental images also have applications in kinesthetic intelligence, which uses bodily imagery to carry out tasks such as playing sports.

Imagery is also useful for some kinds of intelligence tests. Raven's matrices make up a nonverbal intelligence test commonly used in educational settings when the estimation of intelligence should not depend on knowledge of a particular language. Figure 2.3 illustrates a sample problem of identifying the missing ninth pattern. Some people may translate the relevant information into words, but for me it is easier to operate with the visual representations. I work out the image manipulation that produces the third item in the first two rows, then apply this same manipulation to the bottom row. As in the supermarket example, intelligence depends on

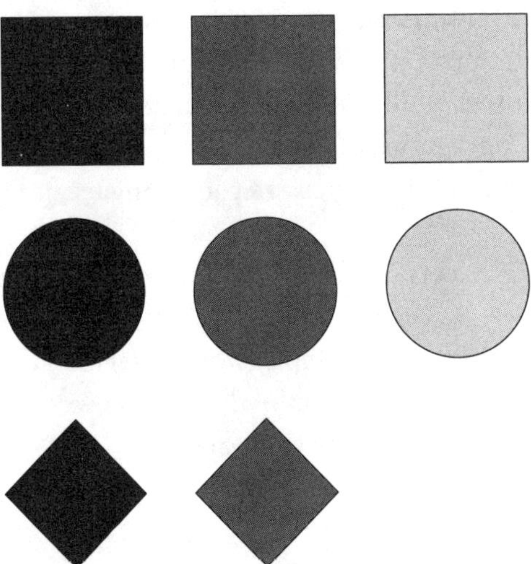

Figure 2.3
Tasks like those in Raven's matrices.

the ability to manipulate mental images. I count imagery as a mechanism, not just as a feature, because it has parts (images and their components) and interactions (e.g., combining, rotating, zooming) that produce regular changes (new images and inferences). In *Brain-Mind,* I show how to translate these mental mechanisms into neural mechanisms.

Concepts

Concepts are mental representations that are often captured by words. For example, in the grocery store you can use concepts such as *coffee, granola,* and *milk,* each of which has a corresponding word. But concepts surpass words by incorporating sensory images such as the smell of coffee, the crunch of granola, and the whiteness of milk. Moreover, you may have some concepts for which you do not have words, such as the smell of blue cheese, which applies to both Gorgonzola and Stilton. Educated people have tens of thousands of words and even more concepts.

Concepts play many roles in intelligent problem solving. You can use concepts such as *lemon* and *lime* to identify similar fruit. By categorizing something as a lime, you permit numerous inferences, such as that it can be squeezed into a gin and tonic. Concepts can also serve as cues for retrieving memories, for example, when thinking of coffee reminds you that you also need coffee filters. Concepts are also crucial for making more complex inferences that involve the rules and analogies discussed later.

Possessing concepts is crucial for answering IQ questions that test verbal knowledge. To answer the question in box 2.1, you need to understand the meaning of the concepts *talkative, job, ecstatic, angry, windy,* and *loquacious* and be able to carry out the exercise of comparing meanings of the pairs of concepts.

Like imagery, concepts count as a mechanism rather than just a feature, because they have well-specified interactions such as combination into more complex concepts, for example, when *natural* and *selection* combine to form *natural selection.* My book *Brain-Mind* provides a full analysis of concepts as mental and neural mechanisms.

Rules

Rules are mental representations that use more than one concept connected by if-then relations, such as "If something is a lemon, then it is yellow." In cognitive science, a rule is any general statement, not just a prescription

Box 2.1
Sample IQ question depending on concepts

Find two words, one from each group, that are the closest in meaning:

Group A

talkative, job, ecstatic

Group B

angry, wind, loquacious

Possible Answers

a. talkative and wind
b. job and angry
c. talkative and loquacious
d. ecstatic and angry

Correct Answer

c. talkative and loquacious

Source: http://examples.yourdictionary.com/examples-of-iq-questions.html

such as "If you have under eight grocery items, then go to the express line." Such rules can chain together series of inferences to solve problems, for example, in the reasoning "That fruit is yellow, so it's not a lime, so I don't want it for my gin and tonic." Rules are also useful for expressing more complicated relations among concepts, for example, in saying that if you want to make a healthy salad, then you can buy lettuce, tomatoes, and cucumber. COVID-19 introduced new rules such as *If you go grocery shopping, then wear a mask.*

Rules can also be useful for generating explanations. If you are thinking of buying some fish, and you notice that the rainbow trout looks a bit slimy, then you can abductively infer the explanation that it is probably old, using the rule that if fish is old, then it gets slimy. Rules can also be useful for making decisions, as in my rule *If I need whitefish for a recipe, and there are different varieties of whitefish available such as red snapper and halibut, then pick the fish that looks freshest.*

Rules are often useful for solving problems on IQ tests. To come up with the answer to the question in box 2.2, you need to know the background rule that says that people age at the same rate, one year per year. You also

Box 2.2
IQ question that requires rules to answer

John, twelve years old, is three times as old as his brother. How old will John be when he is twice as old as his brother?

Possible Answers

15 16 18 20 21

Source: https://www.mensaiqtest.net/mensatest.

need mathematical rules such as that $3\times4=12$ so that you can calculate that John's brother is currently four years old.

Learning to solve problems better requires learning new rules. A few rules may be innate, for example, when newborn babies recognize that milk tastes good. But most rules have to be learned, for example, when someone teaches you that ripe watermelons make a hollow sound when tapped. If you tried several Granny Smith apples and found them sour, then you can generalize that you should not purchase them. More complicated abductive reasoning is required to make inferences about rules that go beyond experience, for example, inferring that a fish is smelly because bacteria are growing on it.

Analogies

Grocery shopping and other problems are easiest to solve when people have general rules that apply to them. But for many new situations, people lack a store of rules and concepts that can generate answers. Analogies enable you to solve a problem based on just one previous case that was not sufficient to generate a rule. In a 2019 discussion on CNN, Chris Cuomo said that President Donald Trump was producing a "lasagna of lies." Cuomo's analogy is that Trump's lying came in many layers, like the layers of pasta, sauce, and cheese in a lasagna. Humans use analogies in diverse applications such as jokes, everyday problem solving, and scientific discovery.

Suppose you have only made one fruit salad in your life without a recipe, but it was a hit with your dinner guests, using strawberries, bananas, apples, oranges, and blueberries. By analogy, you can make the same salad again, but while shopping, you may see that blueberries are not available. You can decide to substitute raspberries for blueberries to produce an analogous salad.

Cognitive scientists have worked out the mental mechanisms that enable people to use analogies right up to advanced forms such as scientific discovery. You face a target problem, such as what to make for dessert. The source problem is a solution that worked in a previous case, your first fruit salad. You need to adapt the source to the target to produce a solution to the target problem. Doing so uses simple concepts such as *fruit*, but also more complex relations such as *same size* and *same approximate sweetness*. Mapping the source problem to the target problem operates with various constraints such as maintaining the structure: you want fruit to go in the bowl, not the bowl in the fruit.

Analogies are often used in intelligence tests, as in figure 2.3, which uses a visual analogy that requires imagery. The Miller Analogies Test consists entirely of analogies. Analogies are intellectually challenging because they require minds to keep the source and target problem simultaneously in working memory, and they depend on considerable knowledge of language and the world. For example, to answer the question in box 2.3, you need to have concepts such as *brother*, *sister*, *niece*, and *nephew* and know rules such as that brothers are male and sisters are female. Moreover, you need to figure out that the most relevant difference between brother and sister is that one is male and the other is female, and then transfer this relation to the concept of niece.

Being good at analogies requires both cultural learning and mental processing power. The analogy problem in box 2.4 requires knowing that composing and painting are both ways of artistic production, and that Monet was a painter. Solving it can only be accomplished by people who have acquired concepts like *musical composition* and have learned about different composers and painters. The problem is also challenging for working memory, because you have to keep in mind the different artists, artistic practices,

Box 2.3
Analogy item used in IQ test

Which one of the five makes the best comparison?

Brother is to sister as niece is to:

Mother Daughter Aunt Uncle Nephew

Source: https://www.mensaiqtest.net/mensatest.

Box 2.4
Analogy item from IQ test

> Bach : Composing :: Monet :
>
> a. painting b. composing c. writing d. orating
>
> Source: https://en.wikipedia.org/wiki/Miller_Analogies_Test.

and the relations among them. If the IQ test or exam is operating with a time limit, then speed of processing also matters. Individual differences result from a combination of educational background and mental capacity.

Very young children have difficulty with analogies, but by around age four, they have the working memory capacity to be able to understand relations among relations, for example, grasping correspondences between fables and real life. Part of getting better at analogies is learning a large supply of cases to serve as source problems that can then be used to suggest new solutions to target problems. If you know many cooking recipes, then you have a large store of sources to adapt to produce new recipes by analogy. You cannot do much to improve your working memory capacity and speed of processing. But educating yourself to have a bigger stock of source analogues and being able to recognize problems that require analogical thinking can make you more effective analogically, and hence more intelligent.

Emotions

Emotions sometimes get in the way of effective grocery shopping, for example, if you get excited about buying large amounts of expensive beef tenderloin. But emotions also contribute to intelligence in important ways. Emotions make six main contributions to intelligent actions: evaluation, information, attention, motivation, memory, and communication.

First, emotions provide an ongoing evaluation of how you are doing. Evaluation of prospective purchases can sometimes be done by numerical cost-benefit calculation, but such calculations are time-consuming and depend on numerical information that is rarely available. In contrast, you can evaluate a product in a grocery store quickly and effectively by having a feeling ranging from enthusiasm to disgust.

Second, emotions provide information about the relevance of possible actions to your goals. You may not even know what your goals are, but if you find yourself really wanting to buy an expensive cheese, then you can realize that you care more about getting tasty food than about saving money.

Third, emotions help to focus attention on what matters to you. I find grocery shopping boring, so my mind tends to wander toward personal and professional topics. But then I get annoyed with myself, and this annoyance helps me to refocus on shopping for the food that I need to provide the energy to pursue more interesting activities.

Fourth, emotions provide motivation to carry out actions. Because I dislike grocery shopping, I need motivation to go to the store, such as the fear of hunger if I run out of food, or the desire to make guests happy. The purely verbal inference that my cupboard is bare and therefore I should make my way to the store does not have as much impact on my actions as states of fear or desire. Emotions also help to form bonds between people that motivate them to live and work together.

Fifth, emotions contribute to memory both by helping to determine what is important enough to remember and by providing cues for recalling relevant events. For example, if I embarrass myself by dropping a jar of pickles in the store, then the intense emotion will make me remember this event and recall it in the future when I need to be more careful with jars.

Sixth, emotions contribute to communication between people concerning their mental states. If I smile genuinely at a cashier while saying thank you, I am conveying that I am happy with the assistance that I have received. On the other hand, if I scowl at a slow and ignorant person at the fish counter, then I am communicating that I am unsatisfied and annoyed with the service.

Current theories of emotions offer different mechanisms for how they work. Some theorists claim that emotions are appraisals, which are judgments made of the extent to which a situation satisfies goals. Other theorists emphasize changes in the body as interpreted by your brain. I prefer a new theory of emotion that describes how the brain integrates both appraisal and physiology while taking into account social context. For shopping, emotional feelings result from a brain mechanism that combines unconscious appraisals of how well a purchase accomplishes your goals and detection of physiological responses to purchases such as heart rate.

Language

As boxes 2.2–2.4 illustrate, linguistic ability is a major part of IQ tests. The justification for having language skills contribute to IQ tests is that language skills are also big predictors of success in education and work. Language is also useful for grocery shopping and many other tasks. In the store, I often read nutrition labels to find out the ingredients, calories, and carbohydrate-to-fiber ratio. Most of the inferences that I make using concepts, rules, and analogies employ language, although images and emotions show that there is more to thought than language.

The language mechanisms used in problem solving and learning combine syntax, semantics, and pragmatics. Syntax concerns the structure of sentences, for example, reflecting the difference between the sentences "Mary wants food" and "Food wants Mary." The mental mechanisms for syntax include ones for parsing sentences into their components and merging components together into new sentences.

Syntax meshes with semantics, which concerns the meanings of words and sentences. For semantics, the two crucial questions are how words can have meaning in relation to other words, and how words can have relations to the world. Word-to-world meaning comes from connections to sense perception, for example, when you learn the meaning of the word "cat" by seeing, touching, and hearing cats in the world. Word-to-word meaning comes from the associations between words and the corresponding concepts in the mind. By learning that cats are mammals, you expand your understanding of the meaning of both the words "cat" and "mammal" through the mental relations between the concepts *cat* and *mammal*. The mental mechanisms for semantics track the parsing and merging operations of syntax to understand and generate meaningful sentences.

Pragmatics concerns context and purpose. Language is often ambiguous, and solving problems on IQ tests or in real life requires identifying what is meant in a particular context, for example, that the bank of a river is different from a financial bank. Linguistic utterances are produced purposefully rather than randomly, and grasping the cognitive or social function of a sentence is an important part of using language intelligently. Mental mechanisms for grasping context and purpose include ones for satisfying constraints among different interpretations by using interactions between representations of words and sentences.

Computer models of language tend to work on syntax first and then add semantics and maybe address some pragmatic issues. What makes the brain brilliant at language is that its parallel processes, carried out by billions of neurons, can manage syntax, semantics, and pragmatics simultaneously.

People can rapidly produce and comprehend complex sentences with effectiveness that far surpasses other animals and current computers.

Intentional Action

I included acting as one of the features of intelligence because of the importance of doing things in the world rather than just in the head. Acting can be done automatically in reflexes such as ducking a thrown rock and in routine behaviors such as walking. What distinguishes intelligent actions from automatic ones is that they result from intentions to act. For example, I go to the grocery store with the intention of buying food while satisfying various constraints about enjoyment, cost, health, and time. What are the mechanisms by which intentions lead to action?

According to commonsense psychology, actions come about because of beliefs, desires, and intentions. If I believe that spinach is healthy, and I desire to be healthy, then I can form the intention to buy spinach, which leads to the action of buying it. Beliefs are what people think about the world, desires are what they want the world to be, and intentions are commitments to carry out actions to bring about desired outcomes. Interactions among representations of beliefs, desires, and intentions lead to the selection of actions.

Peoples' beliefs, desires, and intentions produce actions when their interactions generate activity in the motor cortex of the brain, which instructs the body to act. For example, when I see spinach in the grocery store, my intention, derived from my beliefs and desires, causes my motor cortex to move my arms to pick up the spinach and put it in my cart. Tobias Schröder, Terry Stewart, and I developed a neurocomputational model of how interactions among six brain areas produce such results. We explained beliefs as neural representations of the world, desires as emotions, and intentions as neural processes that combine representations of situations, emotional evaluations of the situations, and the doing of actions.

Intentions do not always lead to actions, because people can be distracted or tempted to do something more appealing. For example, if I am reaching for tofu but a package of expensive and fatty brie cheese catches my eye, I may react by grabbing the brie rather than the despised tofu. More intelligent action requires keeping in mind all relevant goals, such as health and wealth, so that action results from full consideration of the relevant beliefs and desires rather than whims. People vary in the extent to which they can make decisions based on long-term goals rather than immediate temptations, as in the famous experiment when children have to choose between getting a marshmallow immediately or two marshmallows if they can wait.

Consciousness

You might wonder whether consciousness is even a feature of intelligence, let alone a mechanism. I have already described how feelings that include emotions are important for intelligence, and I look to a theory of consciousness to explain how feelings work. People are conscious of bodily experiences such as pain from a stubbed toe, perceptions such as seeing a watermelon, emotions such as feeling happy, and thoughts such as my awareness that I am writing this paragraph. A theory of consciousness should provide mechanisms that explain the occurrence and characteristics of all such experiences.

Consciousness contributes to intelligent actions in at least three ways. First, it focuses attention on factors relevant to making good decisions, for example, when I consciously think that I should buy tofu rather than brie. Consciousness interrupts automatic processing to make people think carefully about a wider range of factors. Such focus is highly motivating in that you are more likely to do something about a sore finger if it consciously hurts rather than being an abstract concern. Emotions usually include conscious awareness, which highlights their contributions to evaluation, information, attention, motivation, memory, and communication.

For example, when I was first working on chapter 5 of this book, I was having trouble getting down to writing, an unusual occurrence, since I rarely get writer's block. When I stopped to ask myself consciously why writing was going so poorly, I quickly realized the answer: chapter 5 was boring! So I scrapped it, along with similar chapters 6, 7, and 8, and came up with an exciting new plan for writing the second part of the book. Without conscious interruption, I might just have continued to be stuck on the boring chapter and bad plan for the book. Consciousness is also valuable as an interruption in more dire circumstances, for example, when the rattle of a snake shifts your attention from whatever else you are doing. Consciousness enables you to step back, be more deliberate, and figure out what is going wrong with standard problem solving.

Second, consciousness is useful for organizing inputs from different senses, for example, when you tie together the look, feel, taste, and smell of an apple. Consciousness can also help to reconcile conflicts among the senses, such as when something that looks like an apple tastes like a pear. Consciousness provides a unified simulation of complex environments that makes it easier to learn about them.

Third, consciousness often helps in social situations. A major way to figure out what is going on in someone else's mind is empathy, when you put yourself in someone else's shoes and experience a similar emotion. Conscious empathy helps us to explain and predict what other people are likely to do. Without consciousness, people would not be able to approximate the feelings of other people and engage in high degrees of social cooperation.

Consciousness can also be valuable for teaching other people. An experienced driver, athlete, or musician may be effective at performing without conscious awareness. But experts who want to convey their knowledge to others may need to step back and become conscious of what they are doing, for example, moving a gearshift in a particular way to teach other people how to do it. Therefore consciousness makes valuable contributions to intelligent thought: goal-relevant focus, merging different senses, empathic understanding, and teaching.

Some theories of consciousness have concerned only mental processes, for example, proposing that what characterizes consciousness is having representations about representations. But in 1994 Francis Crick inaugurated the attempt to explain consciousness by interactions among brain areas such as the claustrum. Current neural theories include Stanislas Dehaene's view that consciousness is information broadcasting within the cortex, Giulio Tononi's claim that consciousness is integration of information, and my own account that consciousness results from competition among neural representations. Much remains to be known about the causes of conscious experiences, but neuroscience is beginning to provide the basics for an explanation of feelings.

Summary of Mechanisms

My theory of intelligence consists of eight mechanisms that provide the explanations for twelve features. Other mental and neural mechanisms may also contribute to intelligence, but these eight provide a good start for the purpose of comparing humans to machines and animals. The best candidate for a ninth mechanism is memory, including storage and retrieval of information about events (episodic memory), general knowledge (semantic memory), and ways of doing things (procedural memory). All of these can be explained by mechanisms connected to images, concepts, and other kinds of mental representation.

Table 2.1 summarizes how these mechanisms consist of connected parts whose interactions produce changes that are important for intelligent behavior. My book *Brain-Mind* shows how all eight of these mental mechanisms can be carried out by a common set of neural mechanisms, many of which have been modeled computationally.

These mechanisms are required to explain the twelve features of intelligence. For example, people solve problems in various ways depending on the nature of their situations and goals using images, concepts, rules, analogies, emotions, and language. Problem situations are sometimes represented in language but can also be depicted in visual or other kinds of

Table 2.1
Summary of mechanisms for intelligence

Mechanism	Parts	Connections	Interactions	Changes
Images	Sensory representations, e.g., visual	Components	Combination, juxtaposition, focus, etc.	Construction of new image inferences
Concepts	Word-like representations	Associations, kinds	Combination, modification,	Classification inferences, formation
Rules	If-then representations	Chains of if and then parts	Deductive, inductive, and abductive inferences	New beliefs and rules
Analogies	Objects, relations, source and target analogues	Binding into structures	Retrieval, mapping, transfer	Use source to solve target problem
Emotions	Situations, feelings	Intensities, valences	Appraisals and physiological changes	Feelings abou situations
Language	Words, sentences	Syntactic structure	Words form sentences; sentences form paragraphs	Utterance, comprehensio
Intentional action	Beliefs, desires, intentions	Representations of situations	Inferences	Actions
Consciousness	Mental representations	Inferences	Competition	Experiences

images, such as navigation problems using a mental map. In familiar problems, you may have an established concept that tells you right away what to do, for example, when you identify a situation as a grocery checkout line and apply your *grocery store* concept.

In more complicated cases, you may need sequences of inferences using rules to figure out a path from your current situation to your goals. If you do not have a store of relevant rules, you may solve a problem by analogy to a previous problem. Emotions contribute because goals come with the desire to accomplish them and fear of failure. Emotions provide inputs to problem solving, but they also provide outputs when solving a problem makes you feel happy or excited. Moreover, emotions accompany problem solving when you are guided by enthusiasm, boredom, anxiety, frustration, or disappointment.

Language contributes to problem solving when goals, situations, and inferences are represented by words and sentences. Language also enables people to abstract away from perceptual representations to generate understanding using hidden causes such as atoms and mental states. Problems that require acting in the world are in part solved by forming intentions that are part of the causes of actions. Finally, consciousness sometimes contributes to problem solving when it provides useful interruptions or social interactions.

Hence the theory of intelligence as multiple mechanisms provides a comprehensive explanation of problem solving, and similar explanations can apply to the other eleven features, such as learning. Such explanations are far more broad, deep, and in line with standards of biological explanation than can be provided by the factor *g* that supposedly explains IQ or the taxonomies of Sternberg and Gardner.

Individual Differences

Why are some people smarter than others, not just in their abilities for grocery shopping and IQ tests but also for their performance in the domains of science, technology, art, and social innovation? Generally, one mechanism works better than another if it has better parts, connections, and interactions that lead to changes that more fully satisfy the purposes of the mechanism.

Accordingly, we should expect people to be smarter if they have more images, concepts, rules, analogies, emotions, language elements, intentions, and conscious experiences. People have huge individual differences in possessing these parts, as in the condition of aphantasia in people who have no mental images at all. Depending on cultural backgrounds, people may have different stores of concepts, rules, and analogies.

The interactions among the parts may be affected by biological factors such as speed of processing and working memory capacity, but the neural basis for these differences is unknown. Individual differences in emotional intelligence may reflect biological differences in physiology and brain structures leading to different emotional responses. For example, psychopaths are people who are incapable of caring for other people and appreciating their emotional distress. But most people are capable of improving their ability to regulate their emotions through psychotherapy and getting a better understanding of other people's emotions through education. One illustration of this potential improvement is the finding that people who read literary fiction rich in emotional situations have greater degrees of empathy than people who merely read action stories.

Individual differences in linguistic capabilities may partly reflect innate variations in people's ability to process language, but most differences more plausibly result from variations in cultural learning. Acquiring a large vocabulary for understanding and producing language takes much time and social interaction. Similarly, syntax more complicated than three-word utterances requires culture experiences of listening to many sentences and gets even better when one reads sophisticated prose. Pragmatic understanding of context and purpose benefits from exposure to diverse situations and constraints.

You might wonder about how smart people really are, given that psychologists such as Daniel Kahneman have identified dozens of cognitive and emotional biases that people routinely display. Moreover, humans have trapped themselves into difficult situations such as global warming, population growth, and the election of powerful leaders who act in their own interests. Nevertheless, the substantial accomplishment of our species in spreading around the world and developing art, science, technology, and government shows that people possess abilities to solve problems and learn that signal intelligence.

Benchmarks for Intelligence

The twelve features and eight mechanisms provide a set of benchmarks for evaluating the intelligence of machines and animals by asking the following questions: To what extent does a particular machine or animal exhibit each of the features of human intelligence? To what extent does a machine or animal employ the mechanisms that explain human intelligence? Considering the different forms of machine intelligence now available, to what extent does their aggregate measure up to the capacities of humans as shown by the full list of features and mechanisms? Identifying gaps can serve to guide research and development to provide machines with these capabilities or to suggest ways of protecting humans from dangerously intelligent machines (chap. 7).

For vivid comparisons, I will grade AI programs and animals for each benchmark on a scale of A, B, C, and F. The grade A means that the bot or beast is virtually equal to humans on that benchmark, so trivially humans get As across the board on our own report card. I ignore the complications that some humans do better than others on some of the benchmarks and that there are respects in which some machines and some animals are superior to humans. For example, computers add numbers better than people, and eagles see better than people.

The grade B means that the bot or beast is missing some aspects of the feature or mechanism found in humans, and the grade C means that many aspects are missing. For example, my description of learning listed seven kinds of learning that people can do, so a machine or animal gets a B if it has most of these, but gets a C if has only a few. The grade F means that the bot or beast completely fails to display the relevant feature or mechanism, while recognizing that machines may eventually be programmed by people or self-programmed to acquire the missing characteristic.

Providing benchmark report cards for millions of species of animals or even hundreds of impressive AI programs would be unbearable. Fortunately, my six smart machines and six smart animals from chapter 1 reveal much about the general state of machine and animal intelligence, so let the comparison begin.

3 Marvelous Machines

Artificial intelligence is already changing your life. If you buy things on Amazon or watch movies on Netflix, then you have received recommendations about more things to buy or more movies to watch. These suggestions are made by recommender systems with machine learning algorithms that use your consumer history to guess what you might like next.

Many people interact with their phones or computers using speech recognition such as Amazon's Alexa, Apple's Siri, Google's Home Assistant, or Microsoft's Cortana. These devices require AI to understand requests and generate responses. On some machines, you can sign on by using face recognition, a technology that is also being used for social surveillance. You may have found Google Translate useful for finding equivalent texts in different languages. Much manufacturing has already been transformed by increasing use of robots. Even if your car lacks new AI features such as Tesla's autopilot, you may navigate using electronic maps that give you directions for the fastest route. Poker players may be dismayed that in 2019 researchers produced a program that beats experts at six-player no-limit hold'em.

Other impacts of AI are imminent. In a decade or two, driverless cars will likely be in widespread use, and military forces will increasingly include killer robots. Natural language interfaces will continue to improve and will play a larger role in how you interact with governments, companies, and doctors. Areas such as science, technology, and professional activities will be enhanced by automated systems for problem solving and learning.

Nevertheless, machine intelligence is not about to surpass human intelligence. The field of artificial intelligence has been around since 1956 and caught my interest in 1978 when I read a fascinating paper by Marvin Minsky about knowledge representation. The last decade has displayed remarkable progress, as illustrated by the six smart machines sketched in chapter

1. In chapter 2, I proposed twenty benchmarks for evaluating how machine and animal intelligence measure up to human intelligence with respect to twelve features and eight mechanisms. For each smart machine, I describe what it does, how it works, and how well it satisfies the benchmarks. These comparisons show how far current AI falls short of human intelligence. If you want to see the overall result, skip ahead to the end of this chapter to view report cards 3.1 and 3.2.

IBM Watson

In 1997, IBM's chess playing program Deep Blue won a six-game match against the world's human chess champion. IBM took on a new challenge in 2004 to produce a computer program that could compete on the television game show *Jeopardy!*. This game requires players to understand clues and generate a relevant question. Producing a program that could compete in *Jeopardy!* required advances in managing large databases of information, understanding natural language, generating plausible answers, and competing in real time. IBM called the program Watson after the company's founder. By 2011, IBM had produced a competitive program and was allowed to compete live against two of the best human *Jeopardy!* players, with Watson winning decisively.

What It Does

IBM realized that the technology used for answering questions on the TV show had many other potential uses in fields such as medicine, business, and law. Watson is now one of the major enterprises pursued by IBM in its plan to supersede its previous success as a producer of hardware. Business applications of Watson have included customer service, risk management, and tax law.

In health, a major Watson initiative is called Watson for Oncology, which is designed to help doctors to improve treatment of patients with cancer. IBM has partnered with one of the world's leading cancer hospitals, Memorial Sloan Kettering, to produce an extensive database of patient treatments and outcomes. Watson also accesses countless pages of information from medical textbooks and journal articles. Ideally, Watson for Oncology is supposed to provide suggestions for medical treatments, but it has been criticized for requiring a huge investment of time and money without documentable improvements in patient care.

Watson is also being used as an adviser for scientific research because of the ability to rapidly search large amounts of research literature. The result is a program called Watson Discovery that has been applied to domains in the life sciences including genetics and biochemistry.

Another field that uses large databases extensively is the law. A program called ROSS that is built on Watson technology is intended to help lawyers by answering questions concerning legal issues. Like Watson's applications to *Jeopardy!* and oncology, ROSS rapidly extracts informative answers from extensive databases.

IBM's most amusing application was Chef Watson, which relied on a database of thousands of recipes to answer queries and generate new recipes. For example, you could ask it to find a recipe that uses chicken, cauliflower, and sesame oil; if it did not retrieve a recipe, it created one. Chef Watson produced recipes that sounded novel and worth making, but the website for Chef Watson is no longer available.

Another impressive IBM application of AI is Project Debater, announced in 2018. Given any topic for debate, the program rapidly accesses millions of documents and processes natural language to generate lists of items that can be viewed as both pros and cons for a particular issue. Most impressively, it generates coherent arguments on whatever side of the debate it is assigned, enabling it to engage effectively with skilled humans.

How It Works

Figure 3.1 shows the steps that Watson used to play *Jeopardy!*. Watson first had to understand a question in the form of a clue, such as "President under whom the United States gave full recognition to Communist China." Watson used algorithms to analyze the question to determine what it was asking. It parsed the grammar of the sentence to identify parts of speech such as subject, predicate, and object.

Next Watson generated hypotheses about answers to the question by consulting a database of millions of documents built out of encyclopedias and other reference sources. For *Jeopardy!*, Watson was not allowed to consult the internet live, but it used its database to generate candidate answers that might include various American presidents such as Richard Nixon, Jimmy Carter, and Ronald Reagan.

To play *Jeopardy!* competitively, Watson needed to have a measure of confidence that its best answer was correct enough to win and to avoid

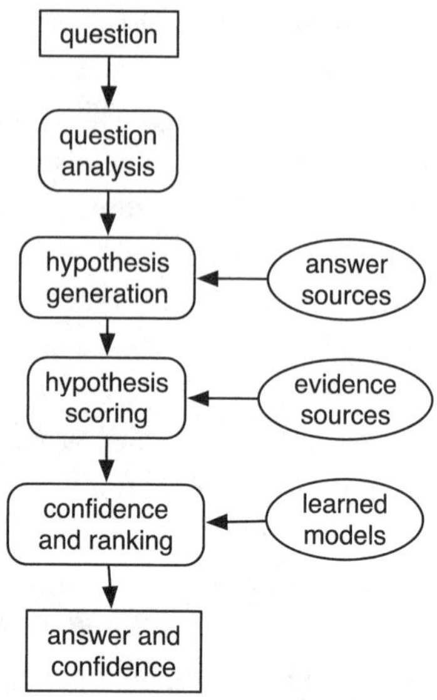

Figure 3.1
How Watson answers questions.

penalized false answers. It scored hypotheses using different classes of evidence, and it used other algorithms to consult answer sources whose word patterns suggest that the answer might be correct. The scoring enabled Watson to rank hypotheses against each other and to assess its confidence in the best answer. This assessment is made statistically based on learning how to weight and combine scores that get better with experience. Putting it all together, Watson produced the correct answer that Jimmy Carter was the president who recognized the communist regime in China.

Jeopardy! requires contestants to answer questions more quickly than competitors, and Watson was too slow running on a single computer. So IBM sped Watson up by having its hundreds of algorithms for finding and evaluating answers run independently on multiple computers. Watson also needed special-purpose algorithms just for playing *Jeopardy!*, such as choosing clues to answer.

IBM researchers have demonstrated that Watson's question-answering capabilities have potential applications beyond *Jeopardy!*. For example, medical diagnosis can be viewed as an attempt to answer questions about the diseases that cause symptoms. Recent versions of Watson extensively use deep learning, which I explain later in the chapter.

Feature Benchmarks

Watson is a major initiative by one of the world's largest computer companies, which touts it as doing "cognitive computing." Watson clearly displays some of the central features of human intelligence. When it answers questions in business, science, medicine, and law, it solves problems based on learning to evaluate possible answers. Watson uses its scoring system for answers to recommend decisions such as what treatment to use for a cancer patient. Project Debater displays the capacity for reasoning in the form of giving reasons for and against a position. Chef Watson shows that the program is capable of creativity by generating recipes that are new, surprising, and somewhat valuable. Watson's ability to take inputs from humans and generate sensible answers displays a rudimentary sort of communicating. So Watson accomplishes many of the features of human intelligence.

But Watson lacks other features. It can process some auditory inputs through its speech-to-text capability, but it has no perceptual abilities to process inputs that are visual, tactile, olfactory, or gustatory. It has no learning capacities for abstracting new concepts from the ones it already has, for example, generating new theoretical ideas that could explain the causes of cancers. The answers it generates are not organized into temporal plans that could capture sequence of actions. It generates only verbal answers and is not capable of performing any physical actions on its own, as it has no robotic interface. Watson has an element of linguistic understanding, shown in its ability to turn linguistic inputs into intelligible outputs, but it does not have the kind of deep causal knowledge needed for understanding why things happen. Finally, Watson does what it does without any aspect of feeling found in sensory experience and emotion.

Mechanism Benchmarks

Watson is even more limited with respect to the eight mental mechanisms responsible for humans' twelve features. Watson works with natural language text in the form of sentences, so it lacks representational structures corresponding to concepts, although its processing of words implicitly gives it some aspects of concepts. More positively, Watson's millions of sentences include if-then structures, so Watson can approximate rules and the inferential procedures associated with them.

Watson has many of the mechanisms found in sophisticated language use, especially the ability to parse and compose sentences using syntactic information. Watson has one kind of semantics, the word-to-word kind, but its lack of perception and action shows that it lacks word-to-world semantics. Some philosophers such as John Searle have claimed that no computer can ever have symbols that meaningfully refer to things in the world, but the self-driving cars discussed later in the chapter refute that claim. Nevertheless, most AI systems, including Watson, are semantically deficient in lacking symbols connected to the world.

From the perspective of human intelligence, Watson is dramatically deficient in not employing mechanisms of imagery, analogy, emotions, intentional actions, and consciousness. Of these, analogy is the mechanism that could most easily be added to Watson, because AI already has computational accounts of analogical mapping and retrieval such as the Structure-Mapping Engine of Ken Forbus and Dedre Gentner.

Evaluation

IBM Watson has both impressive accomplishments and notable limitations. It succeeded dramatically in accomplishing its initial goal of beating top human players on *Jeopardy!*. Chef Watson showed that the program could produce creative recipes. Business applications of Watson are currently operating in areas such as tax return preparation, automobile advice, and banking. The Watson for Oncology project displays near-human expertise in the treatment of numerous kinds of cancer and is being extended to others. Project Debater is impressive in composing arguments on virtually any topic. This range of applications shows that IBM's AI efforts have more generality than is usually found in artificial intelligence programs.

Watson succeeds by virtue of incorporating a few of the mechanisms of intelligence, including rules and some aspects of language and concepts, but it lacks mechanisms for imagery, analogy, emotion, action, and consciousness. With respect to features, Watson is capable of some kinds of problem solving, deciding, learning, reasoning, creativity, and communication. But it falls far short of human performance with respect to perceiving, abstracting, planning, understanding, feeling, and acting.

Some of these limitations could easily be overcome. It should not be hard to interface Watson with robots to make it capable of perceiving and acting. Watson's ability to construct an argument should be generalizable into the capacity for sequential planning. Watson's sophisticated linguistic abilities should be adaptable for analogical reasoning.

Other aspects of intelligence, however, seem far beyond the capabilities of Watson and other AI programs. Watson uses various kinds of learning but is not capable of producing abstract concepts. Although getting Watson to interact with the world robotically would not be too hard, it would require a much bigger leap to make Watson's linguistic representations tightly attuned to sensory and motor inputs and outputs, which would be required for explanatory understanding based on causality. In chapter 5, I discuss the origins of human understanding of causes and what it would take to have similar understanding in machines. Major leaps of theory and programming would be required to make Watson capable of imagery, emotion, and consciousness.

Watson clearly ranks as one of the most impressive approaches to artificial intelligence today, but it has a long way to go to pose a threat to human intelligence. For the foreseeable future, Watson will serve as a supplement and aid to what people can do, not as a transcendent replacement.

DeepMind AlphaZero

Some people like to exercise their intelligence by playing classic games such as chess and Go or video games such as Fortnite. Such games require problem solving, learning, perceiving, and other features of intelligence, so they provide a good test bed for assessing how well machines can approximate human performance. Besides IBM's accomplishments with chess and *Jeopardy!*, the company most identified with game-playing prowess today is DeepMind, founded in London in 2010 and acquired by Google in 2014 for 400 million British pounds (more than a half billion US dollars). DeepMind's website in 2019 announces its aim: "Solve intelligence. Use it to make the world a better place."

What It Does

DeepMind has had substantial success by applying neural network techniques to successively more complicated games. It started by showing that simple computer games such as Pong could be learned by a computer program, and has since had amazing success with more challenging games such as chess, Go, and *StarCraft*. In 2018, DeepMind scored a major breakthrough in the difficult scientific problem of predicting how proteins fold.

DeepMind is expanding its range of applications to include nongaming areas such as health, but its main reputation rests on breakthroughs in games such as chess and Go. Go is an ancient Chinese board game played with white and black stones that are placed on a 19 by 19 grid with 361 squares. This layout allows far more possible moves than occur in chess, which has only sixty-four squares, so it was surprising in 2016 when DeepMind produced a computer program that could beat the highest-ranking player in the world. This program is called AlphaGo because Google established a new parent company, Alphabet. AlphaGo was partly trained by people, but most of its ability came from playing against itself repeatedly and learning to make better moves.

In 2017 DeepMind announced a stronger version, AlphaGo Zero, that was able to beat AlphaGo because it had gotten even better merely by repeatedly playing against itself instead of being trained on games played by humans. DeepMind researchers generalized this program into AlphaZero, a program that was also able to learn how to play chess and shogi at world-champion levels. This program acquired its expertise by playing against itself through the technique of deep reinforcement learning.

How It Works

Understanding how AlphaZero manages to be so successful in playing games requires an appreciation of the power of neural networks, deep learning, and reinforcement learning. Figure 3.2 shows a simple two-layer network that turns perceptual inputs into category outputs. Suppose that the animals found in your neighborhood include squirrels, chipmunks, rabbits, and birds. When you see an animal in the park, you want to recognize it as

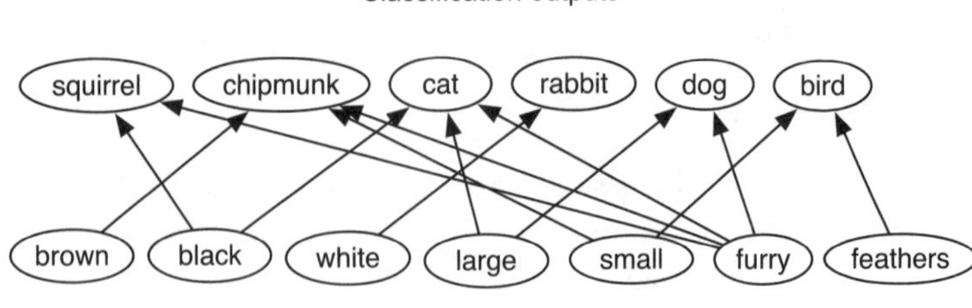

Figure 3.2
Simple neural network that takes perceptual inputs and classifies an object as a kind of animal. Circles represent artificial neurons, and lines indicate connections between neurons.

a particular type. Perceptual inputs to the network indicate that the animal is small or large; brown, black, or white; and quiet or noisy.

These inputs activate the artificial neurons in the input layer, where activation is analogous to the firing activity of real neurons in the brain. Such neurons affect the firing activity of other neurons by either exciting or inhibiting them. Figure 3.2 shows only excitatory connections that lead to some of the neurons in the output layer getting more activation. For example, if the input is that something is small and has feathers, then the output layer will classify it as a bird. It might seem that such networks are merely implementing rules that are easily expressed in language, but they are capable of performing in much more statistically complex ways. A richer network would have inhibitory links between incompatible classifications such as dog and bird.

Children and other people unfamiliar with animals may initially make mistakes and misclassify them. Learning in a neural network consists of changing the strengths of the connections between the neurons, for example, between the input neuron for *large* and the output neuron for *squirrel*. In the 1950s, learning algorithms were developed for the kind of two-layer network shown in figure 3.2, but they proved to be limited in the complexity of the inferences they could handle.

In the 1980s, Geoff Hinton and other researchers showed that three-layer networks like the one in figure 3.3 are much more powerful in their

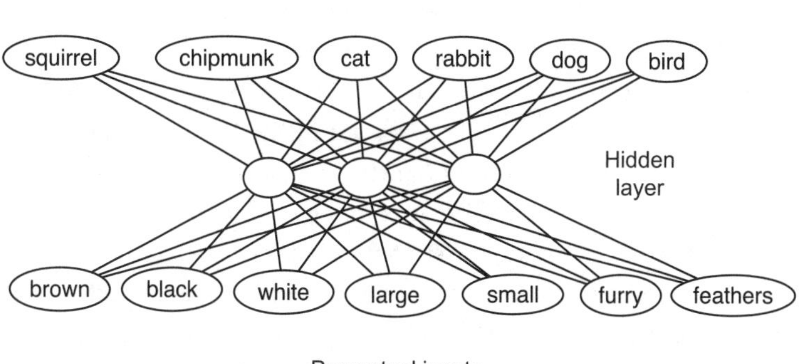

Figure 3.3
Three-layer network capable of learning by propagating errors back through the networks and changing strengths of connections.

ability to learn statistically sophisticated associations between input features and output features. The intermediate layer is called the hidden layer because it is neither input nor output. This kind of network is trained by giving it inputs that spread activation via the hidden layer to the output layer, generating a classification that might be wrong.

Training consists of propagating errors at the output level backward through the network by changing strengths between the layers. For example, if an input that says an animal is large, black, and quiet leads to activation of the neuron for *bird*, then the backpropagation learning algorithm changes all the connection strengths that led to the wrong prediction.

In 2006, Hinton and his students presented new algorithms for training "deep" neural networks with more than one hidden layer. Computers had become faster, allowing for more rapid training, and large databases of inputs and outputs had become available. With more layers, better algorithms, faster computers, and huge training sets, deep learning started to become enormously successful in industrial applications such as recognizing handwriting. In the next decade, deep learning found thousands of different applications, such as face recognition, speech recognition, and translation. Some effective deep neural networks use more than a hundred hidden layers, each with more than a thousand neurons. In 2019 the Turing Award (the top prize in computer science) was awarded to Hinton and fellow deep learning researchers Yoshua Bengio and Yann LeCun. Impressive applications of deep learning include the poker player DeepStack and medical uses such as assessing cardiovascular risk.

DeepMind's breakthroughs came by combining deep learning with reinforcement learning, which rewards effective behavior. Suppose you are trying to train a bird to land on a particular perch. You can reinforce such behavior by a setup that provides a reward in the form of food whenever the bird lands on the perch, training the bird to land on the perch more frequently. In games, reinforcement comes by noticing when moves lead to victory and then changing the strengths of the connections between the neurons to encourage using those moves. DeepMind developed new algorithms, inspired by known brain mechanisms, that use rewards to continuously shape representations in deep networks. The neural networks trained by deep learning can be randomly modified to generate new networks whose effectiveness can then be evaluated by reinforcement learning.

Feature Benchmarks

DeepMind's programs' ability to exceed human performance in video games, Go, chess, and protein folding is astonishing. They excel at particular kinds of problem solving such as figuring out how to make moves in games that require planning and deciding. AlphaZero powerfully combines two important kinds of learning: deep learning that changes connection strengths in multilayered neural networks, and reinforcement learning that uses rewards to learn from successes and failures. Experts in chess and Go have applauded DeepMind's algorithms for their ability to come up with moves that are new, surprising, and valuable, so programs such as AlphaZero qualify as creative. For example, the AlphaZero chess program makes surprising but effective moves such as sacrificing pieces for long-term positional value.

Nevertheless, DeepMind's programs are missing important features of human intelligence. They rely on electronic inputs about board positions rather than perceiving them as a human would, although some robot applications are under way. The deep neural networks are capable of identifying subtle connections between perceptual inputs and classification outputs, but they have no capacity to develop abstract characterizations that go beyond perception. For example, the concept *pet* requires abstractions about ownership of domesticated animals being kept for companionship. DeepMind gets great results by applying associative and reinforcement learning but ignores five other kinds of human learning listed in chapter 2: analogical, causal, hidden-cause, imitation, and teaching.

DeepMind's neural networks are effective in planning and deciding, but they are incapable of stepping back and reasoning about what they are doing or providing causal understanding of their choices. They produce effective computer outputs but do not act on the world. DeepMind's representations are partly based on the human brain, and new applications to multiplayer games introduce teamwork. But the designers make no attempt to introduce feeling into their programs, even though human game playing is full of emotional hunches, worries, and regrets.

Mechanism Benchmarks

DeepMind's programs cleverly use some of the mechanisms that operate in human brains, reflecting the founders' background in computational neuroscience. The fundamental form of representation is not symbolic, based on human language, but rather statistical, distributed across large numbers of neurons. The knowledge in such systems is not based on concatenations of verbal symbols and sentences but instead is located in the connections among neurons. Learning involves not adding new symbols and sentences but rather changing the strengths of the connections, using both deep and reinforcement learning. So DeepMind's programs are somewhat like the human brain and can be interpreted as having concepts, rules, and intentions understood as patterns of firing resulting from connections among thousands of neurons.

However, there are brain mechanisms important for human intelligence that artificial neural network approaches rarely employ. People gain thousands of new neurons every day in a process called neurogenesis, which seems to be important for learning and emotional stability. The kinds of artificial neural networks used by DeepMind have a given stock of neurons and are not regularly generating new ones. The neural networks used for deep learning have excitatory connections between neurons, but real neurons also have inhibitory connections where the firing of one neuron tends to slow down the firing of another. Approximately 20 percent of the connections in human brains are inhibitory rather than excitatory, which is important for making decisions and for controlling the overall activity of the networks. Another limitation of the neural networks used in deep learning is that their computations only use the rate of firing by neurons, ignoring the particular patterns of firing (e.g., fire-fire-rest versus fire-rest-fire) that increase computational power.

Brains are also diverse in the way in which they perform excitation and inhibition, using around one hundred different neurotransmitters that operate with different time signatures and in different pathways. Glutamate is the main neurotransmitter for excitation, and GABA is the main neurotransmitter for inhibition, but the subtleties of brain operations depend on numerous other neurotransmitters and neural modulators such as norepinephrine, dopamine, and serotonin. The firing of neurons in the human brain is affected not only by neural connections and neurotransmitters but also by the chemical signals from glial cells and the circulation in the bloodstream of hormones like cortisol, testosterone, and oxytocin.

From the perspective of psychological mechanisms, DeepMind's programs are even more different from humans. The program's neural networks do not explicitly use concepts and rules, since they have no symbols, but such representations are implicitly captured by the connections in the neural network. For example, a network trained to successfully classify animals and squirrels can be interpreted as having the concept of squirrel even though it has no verbal symbol for it. Neural networks operate mostly by statistical associations, but sometimes these associations can add up to more categorical rules, such as that squirrels have fur. Humans operate with one huge neural network that works for solving problems and learning in countless domains, but DeepMind's networks are trained specially to apply to narrow problems.

The neural networks developed by deep reinforcement learning do not employ imagery. Many people who play games like chess and Go use a visual image of the board and piece location to imagine possible moves and outcomes, but deep reinforcement networks do not produce images. The only neural network system that is capable of working with imagery is Chris Eliasmith's semantic pointer architecture.

Deep neural networks also are incapable of performing analogies, which require relations among relations, including causality. The kinds of neural networks employed by DeepMind work well for noticing associations between concepts, but not for distinguishing between dogs chasing cats and cats chasing dogs. Neural networks in the semantic pointer architecture, on the other hand, can appreciate complex relations that contribute to analogies.

Deep learning has proved effective for some language-related tasks such as speech recognition and translation, but DeepMind has not used either deep learning or reinforcement learning to tackle general problems of language processing that require integrated knowledge of syntax, semantics, and pragmatics. Moreover, DeepMind ignores the contribution of emotions and consciousness to human intelligence.

Evaluation

The productive researchers at DeepMind do not claim that their successful programs offer a general theory of the human mind, although they do imply that they are on the way to "solving" intelligence. But my twenty benchmarks show that DeepMind so far falls well short of human intelligence. Its programs display only seven of the twelve features of human intelligence, and each of these, such as learning, captures only part of human performance. The five missing features are perceiving, communicating, acting, understanding, and feeling, although the first three could fairly easily be added by interfaces with robots.

I described how the neural networks used by DeepMind correspond roughly to human brain mechanisms, but pointed to the absence of important aspects such as neurotransmitters. Programs such as AlphaZero have mechanisms for working with concepts, rules, and intentions but show little potential for handling imagery, analogy, language, emotions, and consciousness. The amazing success of DeepMind in producing programs that surpass humans in games and protein folding encourages overestimation of their approximation to human intelligence. Despite its recent incursions into science and medicine, DeepMind's approach seems less general than the accomplishments of IBM Watson across diverse domains. Both pale beside the generality of human intelligence across tasks that include foraging, agriculture, child-rearing, navigation, technology, science, medicine, and the arts.

Self-Driving Cars

Because I am a full-time writer who works at home, my car mostly stays in my garage. I hope eventually to do without a car altogether when self-driving cars become available that I can summon whenever I need to go somewhere. Driverless cars are already operating in cities such as Phoenix, and major companies are working toward the operation of fully autonomous vehicles.

Cars that drive themselves were only an AI fantasy until 2005, when a vehicle from Stanford University won the DARPA Grand Challenge. In 2009, Google set up a self-driving car project led by Sebastian Thrun, the computer scientist in charge of the 2005 success. Seven years later, Google established a separate company, Waymo, to commercialize driverless cars, which are now being tested on city streets in Phoenix and elsewhere (fig. 3.4). By 2018 their fleet had driven more than ten million miles, in addition to more than seven billion miles in simulation.

Other major companies also have rapidly developing projects for self-driving cars, including Tesla, Ford, General Motors, Mercedes-Benz, and Uber. Tesla has a million cars on the road to provide data to improve its autopilot system, which is currently available to assist drivers but is expected eventually to provide full self-driving capacity. The technology is also being developed for self-driving trucks that can operate in platoons on highways by rapidly communicating with one another. Self-driving cars still have problems dealing with severe weather and high-speed lane changes, but engineers are optimistic that the problems can be handled by better programming, learning, and sensing.

Figure 3.4
Waymo self-driving car.
Source: Wikimedia Commons, licensed under the Creative Commons Attribution-ShareAlike 4.0 International license.

What They Do

According to the World Health Organization, more than a million people die in traffic crashes every year, mostly owing to human mistakes. People are easily distracted by conversation, road events, or emotional problems, and they are often impaired by alcohol, drugs, or sleepiness. In contrast, the computers that operate self-driving vehicles can drive twenty-four hours a day without losing concentration on the road. The elimination of drivers would dramatically reduce the cost of transportation for taxis and long-distance trucks.

Self-driving cars are already capable of performing the functions of human drivers. The first is perception, when cars need to figure out what is going on around them. People mostly use vision for this purpose, although sound is sometimes useful for detecting other cars. Autonomous vehicles have cameras for visual detection of the road, other vehicles, and pedestrians, but they can use other sensors that include GPS, radar, sonar, and lidar.

While perceiving their environments, self-driving cars make plans and decisions about where to go and how to get there. Planning can involve mapping out a route using tools familiar to human drivers, such as Google

Maps. Within that general plan, the car or human driver needs to make constant decisions about going forward, stopping, turning, and speeding up or down. Early driverless cars had to rely on human programming to provide instructions for what to do, but learning based on millions of miles of driving now enables them to make informed decisions.

How They Work

Figure 3.5 depicts how self-driving cars use sensory inputs to make inferences about their environments concerning roads, other vehicles, and potential obstacles. Early autonomous vehicles were poor at such perceptions but got much better when machine learning was introduced. The 2005 winner of the DARPA challenge spent months driving around the desert learning to discriminate between rocks and shadows.

Like the human eye, cameras depend on the reflection of light back to them but, except for flash cameras, do not generate the light. In contrast, radar sends out radio waves, sonar sends out sound waves, and lidar sends out light waves produced by lasers, with reflections and interpretations that identify objects in the environment. Unlike Waymo and most other producers of autonomous vehicles, Tesla does not use lidar and hopes to get by with multiple cameras, radar, and sonar.

Self-driving vehicles integrate multiple signals with past experience to infer the current state of the roads on which they are driving. The vehicles use various algorithms for combining sensory inputs and previous

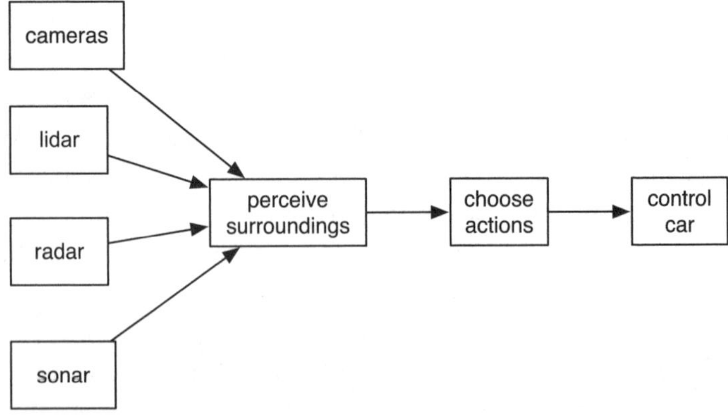

Figure 3.5
How self-driving cars perceive and act.

experience to make inferences. Google's early vehicles used Bayesian networks with variables connected by conditional probabilities. For example, one variable X might represent a car ahead, and another variable Y might represent a sensory pattern. Then the Bayesian network can use information such as the probability of $P(X \mid Y)$ to calculate the probability that there is a car. Later, as Google's expertise with deep learning increased, Waymo cars began to use deep learning in neural networks to make perceptual inferences. Tesla also uses neural networks to integrate multiple sources of sensory information into inferences about the environment.

Unlike humans, who have to learn how to operate in the world individually, self-driving cars can benefit from the experience of all other similar cars. For example, every Waymo car can benefit from the accumulating experience of all other Waymo cars, as well from extensive simulations run on computers not directly interacting with the world.

Given sensory information and the goals of the passengers, driverless cars make inferences about what to do. Such inferences can be made by various algorithms including Bayesian networks and neural networks based on deep learning. The resulting decisions serve to control the speed, braking, and direction of the vehicle.

Feature Benchmarks

Compared to people, self-driving cars are still limited in their ability to deal with unpredictable driving conditions such as snow and accidents. But they are already superior with respect to distraction and impairment, so I hope current shortcomings will be overcome.

Self-driving cars already have many of the features that make intelligent humans capable of driving. Cars accomplish perceiving with cameras analogous to human vision, supplemented by additional technologies not built into humans like radar, sonar, lidar, and GPS. However, cars have no use for the other senses operating in people, such as hearing, smelling, tasting, touching, and pain.

In traveling to destinations, self-driving cars solve problems that include planning, deciding, and autonomous acting. They have an advantage over humans with respect to communication because hundreds of vehicles can be in continuous electronic contact to support efficient flow of traffic.

Each self-driving car learns from its own history, but also from the history of other cars. Tesla has a learning advantage because it already has hundreds of thousands of cars on the road whose records can be compiled. Deep learning and other methods for identifying perceptual patterns are associative, missing out on the other kinds of learning, although it would not be hard to include reinforcement learning in cars. But no plans are afoot to include learning by analogy, causes, hidden causes, or imitation. Autonomous cars communicate with one another, but in ways too simple to qualify as teaching. Associative learning generates some abstractions, but not kinds that are causal, such as acquiring the concept *dangerous driver*, which explains erratic behavior.

Although self-driving cars can already communicate by means of electronic signals, they do not have the linguistic capacities to provide reasons for what they are doing. Unlike human drivers, you cannot ask a car to explain why it just turned left. It is a good thing that driverless vehicles are not allowed to be creative, because moves that are new and surprising may well turn out to be life-threateningly dangerous. Similarly, there is little reason why anyone would want to make a self-driving car capable of having feelings, because a car's reliability advantage comes from avoiding the emotional variability and attention failures that bedevil human drivers.

Mechanism Benchmarks

Autonomous vehicles are rapidly becoming intelligent drivers, but their accomplishments result from mechanisms different from those that support human intelligence. Thanks to technologies not built into humans, such as lidar, driverless cars could exceed humans perceptually. But they have no capacity for imagery, for example, to imagine what a car looks like with a person standing on top. Driverless cars do a weak sort of imagining when they anticipate a pedestrian stepping into the street and react by braking, but the anticipation is based on nonimagistic simulations such as probabilities rather than on running a mental movie with dynamic pictures.

Working with Bayesian and neural networks, self-driving cars have no need for language, so they lack explicit concepts and rules. However, they seem to have concepts and rules implicitly, as shown by their ability to discriminate rocks from shadows and people from cars. They also incorporate nonverbal rules such as *If turning left, then activate the left turn signal.* Driverless vehicles clearly carry out actions and might be said to have an implicit kind of intention when they have an internal representation of getting ready to leave the freeway at the next exit.

Although self-driving cars lack language, they have a kind of semantics not found in Watson and AlphaGo. Driverless cars acquire much of their ability to represent the world and move in it by learning from interactions with the world just as brains do. Therefore the Bayesian or neural networks in such cars have a kind of representation-to-world semantics, although they lack most of the rich representation-to-representation connections that come with language. IBM Watson has powerful language processing that gives it syntax without word-to-world semantics, whereas self-driving cars have semantics without syntax. The full language mechanisms found in humans efficiently integrate semantics with syntax and pragmatics.

Although self-driving cars rely heavily on previous experience gained by themselves and other vehicles, they do not use specific analogies in their planning and decision making. Humans sometimes think as follows: yesterday I took that route to work, and it was fast, so today I will take exactly the same route. Autonomous vehicles learn by smaller bits of experience, not by the whole packages that are employed in analogical thinking.

Driverless cars do an appraisal inference when they figure out what is the best route to take to get around an accident, but their extensive sensory input does not include any representation of their own internal physiology. So there seems to be no prospect of driverless cars ever having mechanisms for emotion. Similarly, there is no behavioral or self-report evidence to suggest that self-driving cars are or will ever be conscious.

Evaluation

The main evaluation question for self-driving vehicles is whether they will ever be good enough to be used extensively on human roads. Technological progress has been sufficient that the answer seems likely to be yes, although I am skeptical of some automakers' claims that full autonomy is only a few years away. I find it hard to believe that a robot car can get around in the vicious Canadian winter, but testing on such difficult conditions is under way.

My interest lies in the broader question of how self-driving cars stack up against human intelligence. In some respects, they seem superior through their capacities for sensory inputs such as lidar, continuous attention, learning from the experience of a whole fleet, and rapid communication with other vehicles. Self-driving cars already display partial features of perceiving, problem solving, learning, planning, deciding, acting, and communicating.

In their own ways, they implement mechanisms of intentional action, concepts, and rules.

Nevertheless, self-driving cars will probably continue to fail in the features of creating, reasoning, understanding, and feeling. They also lack five of the mechanisms that support human intelligence: imagery, analogy, emotions, language, and consciousness. Because driverless cars have no pretensions of general intelligence, it does not matter that they lack these features and mechanisms as long as they continue to improve in their assigned task of driving people around.

Autonomous vehicles are robots, computers with bodies that enable them to operate in the world to perform actions. Electronic autonomous robots go back to the 1940s but have become increasingly powerful in recent years. Nondriving autonomous robots are increasingly becoming available, such as Honda's ASIMO and the animalistic robots being built by Boston Dynamics. Instead of just being programmed by people, robots are increasingly relying on learning to improve their performance. For example, a robot called Baxter can be programmed by moving its arm and repeating a task, a kind of learning by teaching. A few robots such as ASIMO have rudimentary communication abilities but fall far short of human intelligence. Similarly, robots are not designed to understand what they are doing or to be creative about finding new ways to do such things. A rare exception is a robot lab constructed in Wales that not only performed scientific experiments but also generated hypotheses that could lead to new experiments. Overall, however, robots today are even farther from human intelligence than self-driving cars.

Alexa and Other Virtual Assistants

I have no personal use for IBM Watson or DeepMind, and driverless cars are still years away. But intelligent speech recognition already plays a big role in my life. Around 2010, I decided to try the dictation software Dragon Naturally-Speaking because I am a slow typist but a fast talker. It made so many mistakes that it was torturous to use. But I tried it again in 2014 when I was embarking on writing three interrelated books that eventually became my Treatise on Mind and Society. Dragon had improved enough to be useful for first drafts, so I dictated the first version of each book, aided by the fact that by 2016 Dragon had become still better. Much of my writing is now initially produced by dictating to Dragon, although I do all my editing by traditional typing.

I also use speech recognition for emails and text messages produced on my iPad and iPhone, because I hate typing with one finger on a tiny keyboard. I always have to make some corrections but still find that dictation saves me a lot of time. I have also experimented with Woebot, a computer psychotherapist that interacts with users to perform a simple version of cognitive behavioral therapy.

A major use of speech recognition is in virtual assistants such as Amazon's Alexa, Apple's Siri, Microsoft's Cortana, and Google's Home Assistant. Apple adopted Siri in 2010, and I occasionally use it to generate web queries and issue commands to my iPad and iPhone. In 2014, Amazon introduced the smart speaker Echo, controlled by the personal digital assistant called Alexa. I have been using Alexa since 2018 to ask questions such as about the weather but mostly to play music. With tens of millions of Echo speakers sold, Amazon's product is the most widely used application for speech recognition, so I will explain how speech recognition works by discussing Alexa. The commercial uses of speech recognition are rapidly increasing in other directions as well, for example, in controlling televisions and automobiles.

What It Does

Speech recognition by computers began in the 1950s but encountered serious problems. When people talk to computers, there is much variation in background noise, accents, pronunciations, speed, volume, and pitch. To understand what people are saying, the computer needs to translate the sounds produced by their mouths into words. For example, if I ask Alexa to "play Bruno Mars," it needs to convert the sound waves produced by my voice into an internal representation of the instruction to play music. Additional steps are required to use that representation to carry out the appropriate instruction. Similarly, other personal assistance and dictation software has to accomplish the task of reliably turning sounds into words and making inferences about how to react. Early speech recognition systems required much training to identify individual voices, but Alexa has no problems with my voice, even when I am hoarse from a cold.

How It Works

Figure 3.6 shows how Alexa and other virtual assistants carry out tasks such as answering questions, playing music, and controlling your home. The first nontrivial task for Alexa to accomplish is processing sound signals.

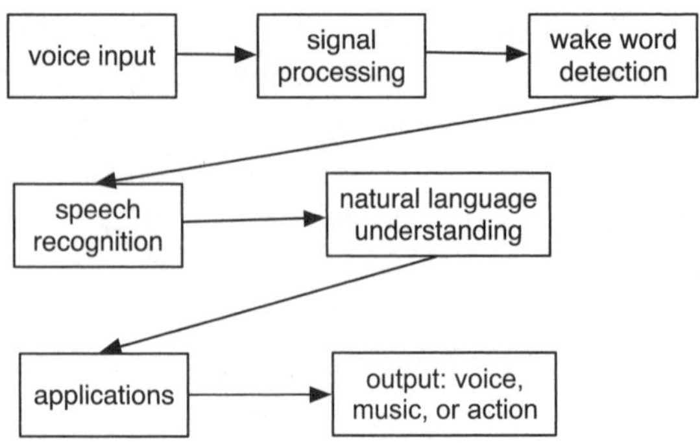

Figure 3.6
How Alexa does what you tell it to do.

Your voice is not the only signal in the environment, which also includes background noises such as furnaces, refrigerators, other people, and music that may be in play, including on Alexa itself. Alexa manages to extract from all these sounds a signal that corresponds to your voice.

The next task that Alexa has to accomplish is to figure out that you are talking to it rather than to yourself or another person. Alexa sorts this out by means of a wake word that you use to address it specifically, which can be "Alexa" or another word that you choose from "echo," "computer," or "Amazon." Signal processing and recognition of the wake word are both performed by the Echo device in your home, but subsequent processing is sent over Wi-Fi to Amazon's computers operating in the cloud. Your home device only needs to do the minimal speech recognition task of noticing that you have used the wake word.

After you say "Alexa," the Echo machine records the signal and sends it to Amazon's computers for much more intensive processing. How does Amazon convert your sounds into recognized speech? This task used to be accomplished by statistical techniques, but most companies now use deep learning neural networks. Around 2010, Geoffrey Hinton, working with Microsoft, Google, and IBM, demonstrated that deep learning does a superior job of speech recognition.

Once Amazon's computers have converted your voice into text, they figure out what that text means. Such inferences require natural language processing of the kind already described with respect to IBM's Watson: parsing string of words into grammatical components and interpreting the sentence. Like Watson, Amazon has no word-to-world semantics, but it has large databases of textual correlations that provide word-to-word semantics. It also has an extensive database of queries that are likely in particular contexts. Amazon is currently restricted to a narrow set of queries, but it is running a contest to reward researchers who can make a virtual assistant carry on a conversation for twenty minutes. As you probably already thought, when Alexa says, "I'm happy you're home," it is not really happy but merely providing a canned response. Similarly, if you tell Alexa that you are lonely, it is only faking when it says, "I'm sorry," although the advice to talk to a friend or go for a walk is well taken.

As soon as Amazon has processed your query, it uses its own resources and the resources of thousands of other companies to do what you want. Alexa can easily access the web to answer questions such as "who is Justin Trudeau?," and it plays music from multiple providers. To carry out other tasks such as controlling your furnace, Amazon needs to use other applications that are called "skills" that can be downloaded and integrated with the Echo. Alexa thereby completes its requested task by producing outputs that can include words, music, or actions in the world.

Feature Benchmarks

Alexa and other virtual assistants accomplish some of the twelve features of human intelligence. Alexa perceives sounds by first using its microphones to receive signals and then using deep neural networks to interpret the signals. Alexa works with Amazon's computers to solve problems that require figuring out what a user wants and then carrying out the requested tasks.

Alexa employs learning at several stages. Training on large databases gives Alexa the ability to recognize wake words and to process natural language to figure out how to interpret the words uttered by speakers. Moreover, Alexa gets better at interpreting the queries of particular speakers by learning to understand their utterances more consistently. Perhaps some abstracting occurs, as deep learning in multilayer networks generates sound categories that do not appear in the initial record inputs, but no theoretical abstraction goes beyond the sensory input. Alexa is communicating by listening to people's queries and responding with answers that are often informative.

Alexa is deciding when it chooses a response to a speaker from among several that might make sense. Alexa sometimes makes bad decisions, for example, when it misinterprets a request for a sports update as a request for a news update, but many of its decisions concerning what to say or do are effective. The results of Alexa's processing of sound amount to acting in the form of producing words, playing music, or changing the environment by modifying lights and furnace activity.

However, there are features of human intelligence that virtual assistants such as Alexa are not designed to accomplish. Alexa produces verbal responses, but none of these consist of reasoning in the sense of providing arguments for conclusions, although perhaps Alexa could incorporate the debating capacity of IBM's Project Debater. Alexa gives individual responses and does not have the capacity to plan a series that amounts to an argumentative conversation. Recognizing speech is not creative because it does not produce anything new, but perhaps Alexa could someday be creative if it develops the ability to carry on a conversation. Alexa does not display any understanding of how it manages to respond to speech. Finally Alexa's voice may sound engaging because it incorporates some of the aspects of human voices, but there is no reason to suppose that Alexa actually has any feelings. Don't fall in love with her.

Mechanism Benchmarks

Alexa and other virtual assistants are sorely lacking in the mechanisms of human intelligence. They do have some language mechanisms such as the ability to go from sound to words and the natural language processing capacity to attach some meaning to those words. Like human speech perception, Alexa accomplishes these goals both by bottom-up processing from sound signals and by top-down expectations based on learning from previous experience. But like AI systems except for robots that learn from interactions with the world, Alexa's semantics are deficient in being only word-to-word.

Alexa lacks the other mechanisms of human intelligence. It has sound inputs but does not store them as images to be manipulated. People can hear a song and imagine it played faster or slower or higher or lower, but Alexa does not retain its sound inputs for future manipulation. Perhaps deep learning produces in Alexa's neural networks something like the abstractions that people use to process sound, but these do not amount to the concepts and rules used by people in other contexts. Alexa's decisions are based on learning from many previous cases, but not on particular analogies. Alexa carries out actions but does not seem to have any internal representations that could amount to intentions for actions. Finally, Alexa does not have any of the cognitive and neural mechanisms of emotions and consciousness, and its verbal expressions of emotion are bogus. In 2019, Amazon patented a version of Alexa that can detect people's emotional states by listening to them.

Evaluation

Speech recognition is an impressive accomplishment of artificial intelligence because Alexa and other virtual assistants succeed in answering people's questions. My Amazon Echo figures out what I am saying to it and responds appropriately by generating most of the answers and music that I want. It achieves its success by carrying out half of the features of human intelligence, including perceiving, problem solving, deciding, learning, acting, and communicating. However, its capacity for abstracting is unclear, and it does not display the features of planning, creating, understanding, reasoning, and feeling.

With respect to mechanisms, current virtual assistants fall well short of human intelligence. They display some language mechanisms for going from sounds to words to responses, but they lack other language mechanisms such as creative generation of novel utterances and word-to-world semantics. The limited roles of virtual assistants today suggest no reason to incorporate the broader mechanisms of human intelligence, including imagery, concepts, rules, analogy, intentional action, emotions, and consciousness. But if such assistants are to accomplish more ambitious goals such as being surrogate friends or effective therapists, then they will require more attention to such mechanisms. For example, one of the most effective tools of friendship and therapy is empathy, which is a kind of emotional analogy in which you put yourself in the shoes of others by imagining how you would feel in their situation.

Google Translate

Today the most commonly used application of artificial intelligence is Google Translate, which helps more than 500 million people every day to navigate among over one hundred different languages. Translation is a complex task that requires intelligence to transform one language into another with approximately the same meaning. Google Translate does not work like a human translator but is sufficiently powerful to be enormously useful.

From the early days of computers in the 1950s, scientists viewed machine translation as a desirable goal, but early attempts failed abysmally. Researchers tried to produce programs that could translate as people do, by breaking down sentences into meaningful components and figuring out how to convert them into sentences in another language. But scientists did not know enough about how humans process language to make this approach effective, and computers were not fast enough to deal with ambiguities such as whether "the pen is in the bank" means "the writing implement is in the financial institution" or "the pig enclosure is beside the river."

In the 1990s, scientists adopted a different approach that gave up on the attempt to make machine translation work like human translators. Instead machine translation took a statistical approach, drawing on databases of previous translations that have been done by agencies such as the European Union and the Canadian Parliament, which operates in both English and French. These statistical approaches began to be useful, and in 1997 the web-based translator Babel Fish became available, although it was ridiculed for peculiar results of retranslation. A decade later, Google came out with its own translation system, which was markedly better than Babel Fish. Google Translate got even better in 2016, when it was revised to use neural networks based on deep learning to produce translations that work on whole sentences rather than just on phrases.

What It Does

Google Translate is easy to use when you go to its website and enter text in any of more than one hundred languages, which it instantly translates into your choice of language. The phone app is even more convenient, because you can dictate the sentences that you want translated, and it will produce a spoken version in the translated language, combining speech recognition and production.

The only language that I know well enough to evaluate Google's translations is French, so let me use Google Translate to put the last paragraph into French:

> Google Traduction est facile à utiliser lorsque vous accédez à son site web et saisissez du texte dans l'une des 100 langues proposées, qu'il traduit instantanément dans votre langue. L'application téléphonique est encore plus pratique car vous

> pouvez dicter les phrases que vous souhaitez traduire et elle produira une version parlée dans la langue traduite, combinant la reconnaissance vocale et la production avec la traduction.

There may be mistakes in this passage that my French is not good enough to detect, but it looks fine to me. Moreover, see what happens when Google translates it back into English:

> Google Translate is easy to use when you visit their website and enter text in one of 100 languages, which they instantly translate into your language. The phone application is even more convenient because you can dictate the phrases you want to translate and it will produce a spoken version in the translated language, combining speech recognition and production with translation.

I see a few deviations from my original first paragraph, but the meaning has survived translation and retranslation. I have been told that Google Translate is not so successful at moving between English and Chinese, which is not as close to English as French is.

Google Translate has imperfections in dealing with linguistic complexities such as ambiguities and gender. It translates "He gave her his ball" as "Il lui a donné sa balle," which is confusing because *lui* as indirect object is gender neutral, and *balle* in French is feminine. The result goes back into English as "He gave him his ball." Because it relies on statistics rather than meaning, Google Translate sometimes goes for the most common interpretation rather than the one that makes most sense in context. For example, "She put the box in the pen" becomes "Elle mit le coffre à jouets dans le stylo," because the most common translation of pen is *stylo*. This makes no sense, because writing pens are much smaller than toy boxes, and the correct term would be *parc d'enfant*. Nevertheless, Google Translate is useful for most purposes.

How It Works

Google's latest technology is Google Neural Machine Translation (GNMT). Like statistical approaches to machine translation, GNMT relies not on humanlike understanding of language but on large numbers of documents that have been translated by humans. It uses these databases by constructing a deep neural network trained on millions of examples to capture the statistical relations between one language and another. Translations between English and French are good because vast numbers of documents are available from the Canadian Parliament and other sources.

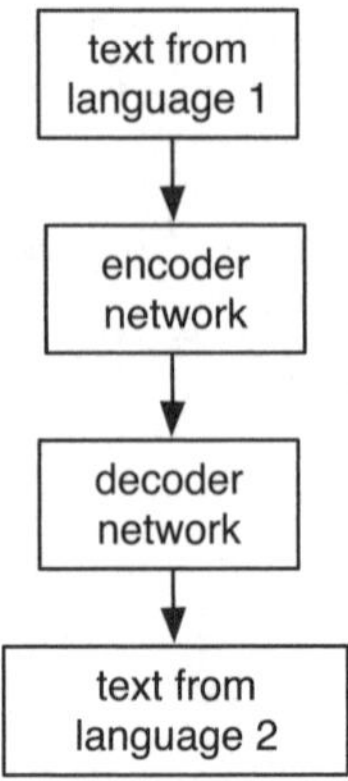

Figure 3.7
Use of neural networks to carry out machine translation.

GNMT consists of encoder and decoder networks, as shown in figure 3.7. The encoder network translates text from the input language into vectors that are long strings of numbers. These vectors are fed into the decoder network to generate predictions about the relevant text in the output language. To speed up computation, multiple computers operate in parallel, rather than relying on the serial processing of single computers.

Feature Benchmarks

Douglas Hofstadter, a famous cognitive scientist and experienced translator, insists that Google Translate operates differently from humans:

> I am not, in short, moving straight from words and phrases in Language A to words and phrases in Language B. Instead, I am unconsciously conjuring up images, scenes, and ideas, dredging up experiences I myself have had (or have read about, or seen in movies, or heard from friends), and only when this nonverbal, imagistic, experiential, mental "halo" has been realized—only when the elusive bubble of meaning is floating in my brain—do I start the process of formulating words and phrases in the target language, and then revising, revising, and revising. This process, mediated via meaning, may sound sluggish, and indeed, in comparison with Google Translate's two or three seconds per page, it certainly is—but it is what any serious human translator does.

We can identify the differences between human translators and machine translators such as GNMT by considering the features and mechanisms of human intelligence.

Google Translate does not do any perceiving, although the phone app version can use speech recognition. Translation programs perform an important kind of problem solving where the goal is to translate from one language to another. Unilingual humans are unable to solve such problems, and few humans can translate between even a handful of languages, whereas Google Translate works with more than one hundred. Its neural networks are based on learning from examples, but the learning is associative rather than causal. Google Translate accomplishes a modest kind of communicating in telling you how to go from text in one language to text in another language.

Although skilled in its narrow purpose, Google Translate shows no signs of other features of human intelligence. It is incapable of reasoning about why it comes up with the translations that it does, and does no abstracting about deeper grammatical features of language such as nouns, verbs, and subordinate clauses. Unlike human translators, Google Translate avoids planning, deciding, or understanding. It is devoid of feeling and has no need for acting on the world. Human translation can sometimes be creative in coming up with original ways to convert one language into another, especially challenging for poetry, where figurative language is paramount. Google Translate relies instead on numerous translations that have already been done by people and therefore is incapable of producing a translation that qualifies as new and surprising as well as valuable.

Mechanism Benchmarks

As my quote from Hofstadter suggests, human translators use all the mechanisms of human intelligence, such as images, concepts, and analogies. Emotion is important when the translator tries to carry over the tone and significance of a text. Human translators also rely on their understanding of human intentional action to capture the meaning of action sequences.

In contrast, Google Translate relies on language mechanisms that are different from the ones that support human natural language processing. GNMT avoids the difficulties of syntax, such as parsing, because it goes from text to vectors to text, using associations learned by deep neural networks. It has a kind of word-to-word semantics that operates across languages, for example, telling you that the French equivalent of "boy" is *garçon*. But Google Translate has no capacity for word-to-world semantics, because it knows nothing about boys and girls in the world. Google Translate does show some pragmatic sensitivity when it considers words in the context of other words, but it has no understanding of the purposes of the texts that it is translating.

Evaluation

Compared to human intelligence, Google Translate is an idiot savant that is precocious at one demanding task but incapable of anything else. It begins to satisfy the benchmarks of problem solving and learning but is borderline on communicating and perceiving. Other intelligence benchmarks are completely lacking, including reasoning, abstracting, planning, deciding, understanding, feeling, acting, and creating.

Although Google Translate uses powerful computational techniques in going from text in one language to text in another, its linguistic capacities are limited by lack of deeper mechanisms for syntax, semantics, and pragmatics. It gets its results by statistical inferences in neural networks without relying on human mechanisms that include imagery, concepts, rules, analogy, emotions, intentional action, and consciousness. Google Translate works much faster and more broadly than human translators but does so by a much more superficial kind of intelligence.

Recommender Systems

When I start Netflix on my television or computer, it provides a list of recommendations of movies or television shows that I might like. The suggestions guide my future viewing, but how does Netflix know my taste? Similarly, Amazon frequently sends me emails concerning things I might want to buy based on my previous purchases. Every day, Apple Music proposes songs I might like based on the ones I have already listened to.

The computer programs that provide suggestions are called recommender systems, and they are used by countless retailers and service providers to let people know about offerings that they might find appealing. The systems qualify as intelligent to a degree because they solve a problem on the basis of learning the preferences of groups of similar people.

What Netflix Does

Worldwide, Netflix has more than 100 million users whose retention as subscribers depends on people liking the movies and television shows that Netflix provides. So Netflix and other providers of goods and services are motivated to suggest products that people will enjoy. When I sign on to Netflix, it informs me of new shows and also makes a set of recommended topics based on my previous choices. If I ever bothered to rate the shows that I have watched, it could also use that information to advise me about similar shows that I might enjoy. Recommender systems such as Amazon rely primarily on a user's previous choices, but the procedure of Netflix is more complicated.

How Netflix Works

As shown in figure 3.8, Netflix uses machine learning to integrate two sorts of information. The first concerns its millions of users, for whom Netflix knows not only what they watched but also when they watched, how long they watched, and what shows they started watching but quit. In addition, Netflix has professional taggers who mark thousands of shows on dimensions such as location, cast, genre, and braininess.

Netflix uses machine learning algorithms to assess what content factors are most important in predicting the watching habits of individuals. From 2006 to 2009, Netflix held a contest with a $1 million prize to see whether anyone could come up with algorithms that could surpass the company's own prediction methods by more than 10 percent. Deep learning from

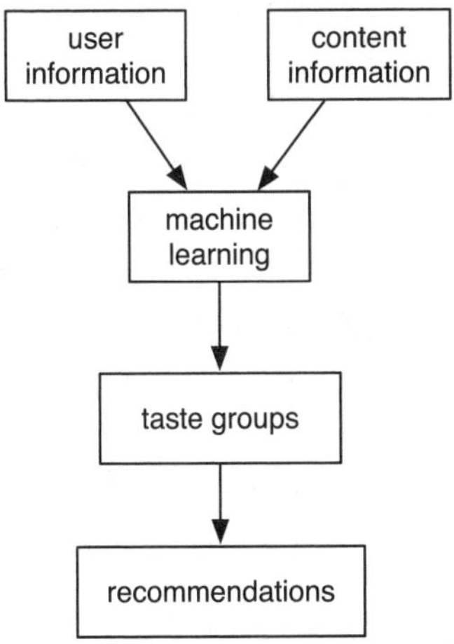

Figure 3.8
How Netflix identifies taste groups that are used to provide recommendations to individuals.

Hinton's Toronto group was one of the finalists, but another group that used a combination of more standard machine learning techniques took the prize.

Netflix's diverse algorithms cluster viewers into more than a thousand taste communities consisting of people with similar preferences. Netflix uses what it knows about you to slot you into a taste community and then make recommendations about what works for that group. Netflix also makes narrower recommendations based on specific shows you have watched. For example, if you watched *House of Cards*, it can provide you with a list of similar shows under the heading "because you watched *House of Cards.*"

Feature Benchmarks

The algorithms used by Netflix predict what grades people will assign to new shows that they watch, so they do help to solve the problem of using information about users and content to make useful recommendations. Moreover, the Netflix recommender system achieves this ability by substantial learning from large databases about shows and preferences and even achieves a degree of abstraction when it organizes watchers into taste communities. Netflix does a kind of deciding when it chooses, from thousands of shows which ones to recommend to a particular individual. These recommendations also demonstrate a weak form of communicating by providing information that people get from their screens.

However, Netflix's recommender system clearly lacks the other features of intelligence. It has no means of perceiving anything or acting in the world, and it does not provide reasoning about its choices. Netflix does not tell you how to plan your evenings, for example, watching a series of shows. Netflix has no understanding of what it is doing through backing its recommendations with causal explanations or reasoning. Netflix's recommendations are meant to provide suggestions that people will quickly recognize and accept, so it makes no attempt at being creative. Netflix aims to elicit positive feelings in its customers, but its algorithms have no feelings themselves.

Mechanism Benchmarks

Although Netflix uses impressive computational mechanisms for learning the factors that predict what shows people like and for clustering people into taste communities, it lacks most of the mechanisms that underlie human intelligence. For people, imagery is relevant to appreciating movies and television shows that rely heavily on visual and auditory representations. My favorite movie is *Casablanca*, and I can easily visualize and empathize with Humphrey Bogart and Ingrid Bergman saying goodbye. But Netflix's content analysis relies on humans to replace such images by verbal tags, throwing imagery away. The tags employ words such as "drama" and "comedy," but Netflix has minimal representations of these concepts and no apparent rules. Netflix does use linguistic expressions to communicate with users, but these are stock templates that operate with no basis in syntax, semantics, or pragmatics. The Netflix recommender system is devoid of intentional action, emotions, and consciousness.

When Netflix suggests that you watch various shows based on ones that you already watched, it uses a simple kind of analogy. The analogies have three gradations, depending on whether they use properties, relations, or causal relations among relations, as the following examples illustrate.

1. *Property analogy*

Movie 1 has a man, a woman, and laughter. You enjoyed it.

Movie 2 has a man, a woman, and laughter.

So you might enjoy movie 2.

2. *Relation analogy*

Movie 1 has a man who likes a woman, and the woman jokes with the man. You enjoyed it.

Movie 2 has a man who likes a woman, and the woman jokes with the man.

So you might enjoy movie 2.

3. *Causal relation analogy*

In movie 1, a man liking a woman causes the woman to joke with the man. You enjoyed it.

In movie 2, a man liking a woman causes the woman to joke with the man.

So you might enjoy movie 2.

Many examples from science and everyday life show that people are capable of using analogies based on causal relations, but current recommender systems have no understanding of relational structures in plots and no comprehension of causality.

Evaluation

Recommender systems by Netflix and other companies qualify as slightly intelligent by displaying some limited aspects of problem solving, deciding, learning, abstracting, and communicating. Other features of intelligence are missing: perceiving, planning, understanding, creating, reasoning, feeling, and acting. Recommender systems are even weaker with respect to mechanisms, as they operate with only weak forms of concepts and analogy, and with no forms of other mechanisms of intelligence. The questions are open whether recommender systems would work better if they became more humanlike, for example, by using richer analogies, and whether such improvements would be cost-effective.

Machine Report Cards

I have discussed only six examples of machine intelligence, which I chose because they are both impressive and popular. I could have analyzed other examples, such as the use of face recognition technology by Facebook and other companies. Moreover, substantial AI research efforts are operating at universities, and the notes to this chapter list some of my favorites. Nevertheless, the six bots discussed here provide an informative sample of what machines can currently accomplish, and show that research is far from achieving human-level intelligence, also called artificial general intelligence.

Report cards 3.1 and 3.2 summarize my evaluations by giving grades to each of the AI systems with respect to the twenty feature and mechanism benchmarks. Following American-style grading practices, the grades are A, B, C, and F. None of the candidates gets an A, which would indicate that the machine comes close to human performance. The B grade means that the system has much of the performance of humans, and a C indicates only a small amount of what humans can do. The F grade indicates failure to even begin to satisfy the benchmark.

Report card 3.1 summarizes how the six machines do against the twelve feature benchmarks. All machines are capable of some basic kinds of problem solving and learning, although they fall short of humans in many ways. The two striking gaps occur in the features of feeling and understanding. All the other features have some of the machines capable of carrying them out. We can conclude that machine intelligence is now approximating some features of human intelligence but has far to go to match it. The

port Card 3.1
aluation of machines with respect to features

ature	Watson	AlphaZero	Self-driving cars	Virtual assistants	Google Translate	Recommender systems
ceiving	C	F	B	C	F	F
oblem solving	B	B	C	C	C	C
nning	C	C	C	F	F	F
ciding	B	B	C	C	F	C
derstanding	C	F	F	F	F	F
arning	C	C	C	C	C	C
stracting	F	C	C	F	F	C
eating	C	C	F	F	F	F
asoning	B	C	F	F	F	F
eling	F	F	F	F	F	F
mmunicating	C	F	B	B	C	C
ting	F	F	B	C	F	F

e text for the grading scheme.

gap between human and machine intelligence is even more dramatic when we recognize that most humans can operate in most of the domains of the six machines, whereas AI systems tend to be special purpose.

Report card 3.2 shows an even sparser accomplishment of intelligent mechanisms across the six machines. None of the six machines implements mechanisms for imagery, emotion, and consciousness. AI systems show some success in dealing with concepts and rules but are just getting started on full mechanisms for language and intentional action.

AI experts disagree about how quickly artificial intelligence can catch up to human intelligence. In 2017 a survey of AI experts found an aggregate estimate of a 50 percent chance of achieving human-level intelligence in machines within forty-five years, suggesting that the boxes in report cards 3.1 and 3.2 could mostly be filled in with As by 2062. In chapter 5, I provide more reasons to be skeptical by listing many advantages that people have over both animals and current machines.

The hardware argument for full AI says: Because of Moore's law that computer speed and memory double every two years, artificial intelligence will soon surpass human intelligence. But physical limitations on the size of computer components imply that past spectacular improvements in computer

Report Card 3.2
Evaluation of machines with respect to mechanisms

Mechanism	Watson	AlphaZero	Self-driving cars	Virtual assistants	Google Translate	Recommen systems
Images	F	F	F	F	F	F
Concepts	C	C	C	C	F	C
Rules	C	C	C	C	F	F
Analogies	F	F	F	F	F	C
Emotions	F	F	F	F	F	F
Language	C	F	F	F	C	F
Intentional action	F	C	C	F	F	F
Consciousness	F	F	F	F	F	F

speed and memory will not continue at the same exponential pace. Moreover, intelligence is a matter of software as much as hardware, and we have no reason to believe that the development of intelligent software follows a pattern of exponential growth. Deep learning with smarter algorithms, faster computers, and larger databases has made great contributions in the past decade, but I have identified kinds of problem solving and learning that present additional challenges concerning causality and other issues.

Arguments for the complete impossibility of human-level artificial intelligence are inconclusive, but enormous technological barriers to producing it persist. I expect incremental progress in filling in the gaps in my benchmark tables, but this enterprise is more likely to take a matter of centuries rather than decades. In chapter 7, I discuss ethical arguments for preventing the AI enterprise from fully succeeding, and I propose steps we can take to limit its progress. For now, the Attribution Procedure from chapter 1 allows us to ascribe some mental characteristics to machines, such as problem solving and learning, but blocks ascription of other important aspects of human intelligence, such as imagery, emotion, and consciousness.

4 Amazing Animals

Which is smarter, a computer or a dog? Simple organisms such as bacteria, plants, and jellyfish have traces of intelligence in their abilities to sense and react to their environments. But I am interested in brainier animals and focus on bees, octopuses, ravens, dogs, dolphins, and chimpanzees. These six beasts provide a good overview of intelligence in nonhuman animals, covering insects, mollusks, birds, and mammals. I overcome the neglect of fish in chapter 6 with my discussion of whether fish feel pain.

For each of the six animals, I describe what they do and how they work using mental and neural mechanisms. This information answers the same questions that I posed for machines: how do these animals rate on the intelligence benchmarks for twelve features and eight mechanisms? At the end of the chapter, report cards 4.1 and 4.2 provide concise comparisons with human and machine intelligence.

Bees

Insects evolved more than 400 million years ago and are not usually recognized as smart. But social insects such as termites, ants, wasps, and bees demonstrate surprising abilities that enable them to solve problems, learn, and communicate with one another. Bees evolved from wasps more than 100 million years ago, adapting to feed just on nectar and pollen from flowers (also newly evolved), unlike the carnivorous wasps. Thousands of species of bees exist, but the honeybee gets most of the attention of researchers.

What They Do

I love honey even though it comes from bee vomit. Honeybees gather nectar from hundreds of flowers and store it in a special stomach where enzymes start to digest the nectar into simple sugars. To continue this process, honeybees regurgitate the resulting mixture into the mouths of other honeybees, where the process of digestion continues. Eventually the result is vomited onto a surface where the beating of bees' wings dries the substance enough that it starts to look like honey.

Much of what bees do is instinctual, for example, slurping up nectar and passing it on to other bees. But searching for flowers requires enough problem solving and learning that there are researchers who study what they call "bee cognition." Bees need to be smart enough to find nectar in changing environments and get back to their hives. Their eyes provide a visual sense that enables them to recognize colors and shapes, and their antennae provide a sense of smell that enables them to detect the scents of flowers and the pheromones of other bees. Honeybees also detect taste through antennae and other body parts that also have receptors for touch. Bumblebees can experience an object using one sense (vision or touch) and then recognize it using the other. Whether honeybees have nociceptors that generate pain is hard to determine, but they have an electromagnetic sense that people lack. Overall, with respect to perceiving, bees rank close to humans, unlike the machines in the last chapter.

Beehives typically contain thousands of bees that forage for nectar. Honeybees can go out searching for flowers over a territory of a kilometer or more, and a day's work can involve visiting hundreds of different flowers. Although bees may only live a few weeks, they get better at solving the problem of finding nectar by learning about their environments and by better discriminating among flowers. They can also learn new techniques for obtaining nectar from particular kinds of flowers.

Bee cognition researchers have investigated the learning capabilities of honeybees in rigorous experiments. Given sugar rewards, bees can be trained to make subtle discriminations using their various senses. They can accomplish planning by efficient foraging sequences that enable them to visit familiar flowers. Bees even have a rudimentary appreciation of number that discriminates zero to three items.

Honeybees are able to communicate to other hive members the direction of good sources of nectar by a waggle dance where their body movements indicate where to go with respect to the direction of the sun. Thus bees learn how to collect food better both individually and collectively as they share successes.

In addition to gathering nectar, the major problem that honeybees solve is finding a location for a new hive. Scouts go out looking for promising locations and use waggle dances to communicate to bees back in the hive where to explore. Through complex interactions, the bees collectively decide on the most promising location and swarm away to establish it.

Honeybees are one of almost twenty thousand species of bees, but the only other species that has been extensively investigated is the bumblebee. Astonishingly, particularly clever bumblebees can learn to get sugar by pulling on a string, and other bees acquire the skill by observing them. This experiment suggests that bees are capable of both tool use and simple cultural transmission based on imitation learning. But the evidence does not suffice to determine whether bees are doing causal rather than just associative learning. Researchers have even claimed that bumblebees are capable of having emotion-like states when they get unexpected rewards, possibly suggesting that bumblebees may have feelings. But experiments are not precise enough to rule out the alternative hypotheses that bees are merely processing rewards rather than feeling them.

How They Work

The operation of machines discussed in the last chapter has little mystery, because their hardware and software are built by human beings. So the mechanisms responsible for the various features of intelligence that they accomplish are easily identified: we know the parts, connections, and interactions that produce smart changes. In contrast, understanding animal intelligence is challenging because it requires figuring out how their brains work.

The brains of bees are much smaller and simpler than those of humans. The volume of the brain of a honeybee is only about a cubic millimeter, smaller than a sunflower seed; and it has fewer than a million neurons, in contrast to the eighty-six billion neurons found in human brains. Honeybee brains are organized into functional areas different from ones found in the brains of humans and other mammals. Bee brains use some of the same neurotransmitters as mammalian brains, such as dopamine and serotonin, but they also use octopamine, which is common in invertebrates and plays some of the same functional roles as norepinephrine in mammals. Nevertheless, miniature brain scans and other technologies are making it possible to figure out how the intelligence of bees derives from neural activity.

The neural mechanisms that support the powerful perceptual capacities of bees are increasingly well understood. Their brains have areas dedicated to processing signals for vision, smell, taste, touch, and a form of hearing. These signals come from the bees' eyes and their antennae, as well as other body parts. As in humans, such signals generate patterns of firing in groups of neurons that can generate additional patterns that represents objects in the world such as flowers with their colors, shapes, and odors. Bees have special brain structures, called "mushroom bodies" because of their shape, that enable bees to integrate different senses.

Researchers disagree about whether bees are capable of image-like representations. Lars Chittka argued that "there is little direct evidence that bees actually perceive images—in the form of little virtual pictures somewhere in the bees' brain." But a small study reported that bumblebees can recognize rotated patterns. James Gould found evidence that honeybees use cognitive maps to navigate with familiar landmarks. In humans, cognitive maps go with visual imagery, for example, when you make a mental picture of your route from home to work or school. Many other animals, such as rats in mazes, operate with cognitive maps, but it is hard to say whether these generate conscious images without additional evidence like the experiments that support the existence of imagery in humans.

Bee neural mechanisms for learning are similar to ones operating in mammals. Learning in neural networks, both artificial and natural, primarily consists of changing the connections between neurons. A simple way this can operate was identified by the Canadian neuroscientist Donald Hebb, who proposed in 1949 that neurons that are connected and firing at the same time have their synaptic connections strengthened. Hebbian learning provides a powerful general mechanism for learning to associate different features, for example, when neurons that represent the color of a flower fire at the same time as neurons that represent its odor.

Bee brains also perform reinforcement learning, similarly to how AlphaZero learned to play Go. Honeybee brains contain a unique neuron, VUMmx1, that responds to sucrose but can learn to respond to an odor if it is followed by a sucrose reward. Learning operates with the neurotransmitter octopamine in a way analogous to how dopamine supports reinforcement learning in mammals. Researchers who think that bees possess emotions also point to these neurotransmitters as part of neural mechanisms suggested to be analogous to human optimism and pessimism.

These learning mechanisms implemented in neurons make honeybees capable of sorting objects into categories such as different kinds of flowers. In humans, concepts usually go with specific words associated with things, for example, the word "geranium" that English speakers use to represent a kind of flower. In brains, concepts are naturally understood as patterns of firing in groups of neurons that incorporate representations of standard examples and typical features.

Research on bees suggests that they have concepts representing not only simple properties such as color and smell but also relations such as *above*, *below*, *same*, and *different*. Aurore Avarguès-Weber and her colleagues report studies that support this interpretation:

> Mastering a relational concept involves encoding a relationship by the brain independently of the physical objects linked by the relation and is, therefore, consistent with abstraction capacities. Processing several concepts at a time presupposes an even more cognitive sophistication that is not expected in an invertebrate. We found that the miniature brains of honey bees rapidly learn to master two abstract concepts simultaneously, one based on spatial relationships (above/below and right/left) and another based on the perception of difference. Bees that learned to classify visual targets by using this dual concept transferred their choices to unknown stimuli that offered a best match in terms of dual-concept availability: their components presented the appropriate spatial relationship and differed from one another.

So it appears that bees operate with the mental mechanism of concepts implemented by neurons. Similarly, bee researchers often describe bees as possessing rules that capture the regularities that they have learned, such as associations between objects, places, and rewards: if a flower has a particular shape and color, then it has nectar.

The communication by which bees inform one another about promising nectar sources requires both a social and a neural mechanism. Social mechanisms are ones where the parts are individual organisms and the interactions consist of transferring information from one organism to another. I always thought that bees interpret the waggle dance of other bees by watching them in a form of visual communication, as when people observe each other's gestures. But the actual form of communication seems to be electromagnetic, as bees emit electric fields during the waggle dance, and other bees sense these fields using their antennae. Neurons in the receptive bees' auditory centers respond by generating firing patterns that represent the direction to fly in pursuit of food.

Feature Benchmarks

Despite their tiny bodies and brains, bees satisfy most feature benchmarks. They are capable of perceiving across the familiar modalities of vision, hearing, smell, taste, and touch and add a sensitivity to electromagnetic fields that is beyond the capabilities of humans. Bees not only observe their environments but react to them through many forms of acting, including walking, flying, eating, and mating.

Bees carry out kinds of problem solving and learning that take them well beyond instinctual behaviors. They can come up with new solutions to problems about finding food and hive locations through extensive exploration, learning, and interaction with other bees. They even seem to be capable of learning of abstract relations such as *same* that go beyond simple perceptual features.

Bees accomplish planning when they work out nonrandom strategies for foraging for plants. Hives engage in collective deciding when they choose locations for new hives. The waggle dance used to suggest hive locations as well as food locations is an effective nonlinguistic form of communicating.

The scientific jury is still out concerning whether bees have feelings such as pain and emotions. In chapter 6, I use the Attribution Procedure to determine whether fish have pain and dogs have jealousy, considering alternative explanations such as motivational drives of the sort found in robots. Sometimes the available behavioral and mechanistic information is insufficient to give a plausible answer

Several features of intelligence are beyond the capabilities of bees. They make inferences about promising food sources but lack the linguistic capacity to provide reasons for what they are doing. Similarly, although they learn causal associations such as what flowers bring rewarding nectar, they do not have any causal understanding to explain what they are learning. Similarly, their learning mechanisms are not powerful enough for creating any representations that are new, surprising, and valuable. The bumblebees learning to pull on strings to get food is not creative because they have to be shown how to do it by humans and then by each other.

Mechanism Benchmarks

Among my eight mental mechanisms for human intelligence, the ones that most clearly lie within the capabilities of bees are concepts and rules understood as patterns of neural firing. Imagery is debatable without more evidence about bee navigation and clearer standards for attributing imagery to animals. Bees can detect similarities and even make abstract relational judgments of same and different, but they show no capacity for analogy.

If bees have emotions, as some researchers suggest, then we could attribute to them some degree of consciousness. However, unlike fish, whose behavior is affected by painkillers, bees do not find relief in morphine when damaged, so it is hard to say that they consciously experience pain. Similarly, their foraging behavior may better be explained by simple, nonfelt reward mechanisms than by emotions that require appraisals as well as physiological responses.

The actions of bees are driven by simple kinds of learning and not by forming the conscious intentions important for humans. Despite their impressive communicative abilities with the waggle dance, bees lack the mechanisms of human language required for processing much more complex syntactic, semantic, and pragmatic information.

Evaluation

Overall, bees are much more worthy of being deemed intelligent than bacteria, plants, or jellyfish. They display eight of the twelve features of intelligence, with the notable absences of reasoning, understanding, and creating, and the possible absence of feeling.

With respect to mechanisms, the most plausible contributors to bee intelligence are concepts and rules, but missing are analogy, intentional action, and language. It is hard to say whether bees have imagery, emotions, and consciousness. Nevertheless, bee behaviors such as foraging, hive creation, and communication mark them as intelligent.

Octopuses

Mollusks such as snails and slugs are not noted for their intelligence. But of the eighty-five thousand species of mollusks, the cephalopods (squid, cuttlefish, and octopus) stand out with respect to large brains and complex behaviors. The octopus is recognized as the most intelligent of invertebrates, far surpassing insects, worms, and other mollusks.

What They Do

Octopuses have powerful senses that enable them to solve the survival problems of finding food and avoiding prey. Their eyes work well in both light and dark conditions and surpass humans in their ability to see polarized light. Their eight arms are covered with hundreds of suckers that accomplish the senses of touch and taste. Chemoreceptors in the suckers are so sensitive that they can pick up chemical information from long distances. In addition, octopuses have a sense of smell through organs at the end of their arms, but octopuses do not have any organs for hearing.

Octopuses rapidly learn to solve complex problems concerning feeding, fleeing, and navigation. The giant Pacific octopus decides between several techniques for eating clams: pulling the shells apart, chipping them with their beaks, or drilling a hole in them. Octopuses adapt to different situations, including the species of clam, shell thickness, and the clam's muscle strength. Octopuses are able to unscrew the lids of jars to get edible ingredients.

To avoid predators, octopuses can move by jet propulsion, squirting water through a tube in their bodies. They can also hide from predators by changing color, pattern, and texture in less than a second. Octopuses are notorious for their ability to escape from aquariums by squeezing through tiny holes or by climbing over walls. One species, the coconut octopus, has even been observed using coconut shells as tools. These accomplishments suggest creative planning, but octopuses show no signs of reasoning about what they do or understanding it by causal explanations. Octopuses can discriminate between different predators, but I have not found any experiments that show them capable of forming abstract concepts like *same*.

Octopuses show shifts in attention based on the availability of food and predators, but whether these shifts are a sign of consciousness is hard to tell. The 2012 Cambridge Declaration on Consciousness included octopuses in the list of animals it took to be conscious, but evidence is limited. It is plausible that octopuses do feel pain in much the same way as mammals and birds, because they have nociceptors and opioid receptors and respond to analgesics. They do not hesitate to nip off their own arms to escape predators, but then guard the injury, perhaps because they are trying to avoid additional pain.

Octopuses are rarely sociable, getting together only to mate in a rather violent fashion, so they have no need for communication. Reports suggest that dosing with the drug ecstasy (MDMA) makes octopuses more sociable, but the drug may merely be interfering with detection of chemical cues of potential mates. Unlike humans, social cognition does not seem to have been a factor in the development of the large brains of the octopus. However, settlements of more than ten octopuses have recently been found, so perhaps they will evolve to become more sociable.

The octopus is probably not the smartest creature in the sea. Among fish, manta rays have the largest brains and may even be capable of recognizing themselves in mirrors. Dolphins and whales are mammals with many more neurons and more complex behavior than octopuses, which nevertheless deserve recognition for showing that intelligence does not require a backbone.

How They Work

The brain of an octopus is only about the size of a walnut, but octopuses have around 500 million neurons. More than half of these are in their arms, which have a large degree of independent control. The base of each arm has a large ganglion (clump of neurons), which leads some scientists to say that octopuses have nine brains to go along with their three hearts, but the heart claim is more plausible than the brain claim.

The octopus has five hundred times more neurons than the honeybee and is comparable in behavior and number of neurons to small mammals such as cats. But its brain is organized differently from mammals, with a vertical lobe system similar to the mushroom bodies found in insects. Nevertheless, octopus neurons are capable of associative and reinforcement learning by changing synaptic connection strengths, just like bees and mammals.

Like bees, octopuses lack the language-oriented concepts of humans but are able to make sophisticated discriminations. Octopuses distinguish among different kinds of predators and respond with different coloring strategies. Similarly, they distinguish among different foods, as shown by their different strategies for opening shells. In humans, concepts serve a variety of nonlinguistic functions by placing objects in the world into useful categories that serve the purposes of action. Octopuses seem to have these kinds of concepts.

It also seems legitimate to attribute rules to octopuses in the sense of if-then patterns of neural representation. For example, octopuses have been known to take a dislike to some of their tenders, adopting a practice that fits the following rule: if that person comes by, then squirt him or her. No one has tried to see if octopuses can learn by analogy, although experiments have shown them to be capable of imitation learning, where one octopus learns to get prey by observing another.

It is difficult to answer the question of whether octopuses have conscious experiences. Peter Godfrey-Smith points out that if octopuses do have experiences, they are very different from those of humans. You are used to the feeling of controlling your arms and feet, but the octopus delegates its motor control to the large number of neurons in its arms. I have found no experiments that address the question of imagery and cognitive maps in cephalopods. Given the enormous differences between octopus brains and human brains, and the lack of identifiable emotion behaviors in octopuses, we have no grounds for attributing emotion mechanisms to them.

Feature Benchmarks

Aside from hearing, the perceiving abilities of octopuses are impressive. They may be color-blind because of limited photoreceptors, but they have a chemical sensing ability not found in humans. The associative and reinforcement learning capabilities of octopuses make them effective at solving problems that involve getting food, avoiding predators, and navigating new environments. Octopuses have some capacities for planning and are capable of acting through moving and operating their arms. They make decisions but lack abstracting, reasoning, and explanatory understanding.

Because octopuses are asocial, they have no mechanisms for communicating with one another, in contrast to waggling in bees, bodily and facial movements in mammals, and language in humans. The first time that an octopus unscrews a jar to get food is new, surprising, and valuable for it and therefore qualifies as creative. Pain behavior and physiology suggest a capacity for feeling.

Mechanism Benchmarks

Although their brains are different from ours, octopuses operate with many of the same mechanisms that make humans smart. They perform inferences and actions by means of patterns of neural firing, and they learn by changing the strengths of the synaptic connections between neurons.

The main mental mechanisms for which we have strong evidence in octopuses are concepts and rules. Learning to discriminate between different kinds of food and different kinds of predators suggests the operations of concepts along with rule-like representations for generalizations and behaviors. Given the series of actions that go into octopus planning, it would not be ridiculous to suppose them capable of intentional actions.

In contrast, I have been unable to find evidence that the mental mechanisms of octopuses include imagery and emotions. Octopuses are often described as being curious, which is an emotion, but a purely behavioral description is that they are prone to explore their environments and to investigate novel objects. I am unsure about whether octopuses have the ability to distinguish same from different, but the analogical inference that involves similarities and differences among relations is beyond them. Although some romantic researchers want to describe emotional consciousness to octopuses, no one supposes that they have mechanisms for linguistic communication.

Evaluation

The behaviors of octopuses support their reputation as the smartest mollusk on the planet, as they satisfy to some degree eight of my twelve benchmarks. That score makes them similar to bees, although bees are better at communicating. Further experiments may show that octopuses are also capable of abstracting and feeling pain.

With respect to mechanism benchmarks, octopuses, like bees, do worse than they do on features. Both kinds of animals seem to have concepts, rules, and intentional actions. Imagery, emotions, and consciousness are possibilities awaiting further evidence. An even longer stretch is supposing that octopuses manage their intelligent behaviors using analogies and language.

It is amazing that a mollusk shows intelligence comparable to that of birds like ravens and mammals like dogs. Nevertheless, with respect to several features and many mechanisms, octopuses fall far short of human intelligence.

Ravens

What are the smartest birds? Honeybees are arguably the smartest insects, octopuses the smartest mollusks, manta rays the smartest fish, and humans the smartest mammals. The scientific consensus is that the brainiest bird families are the parrots and the corvids, a group of 120 species that include various kinds of crows, jays, jackdaws, magpies, and ravens. In problem solving ability, corvids are comparable to the great apes such as chimpanzees.

Among the corvids, ravens have the biggest brains, with more than two billion neurons. Ravens are large black birds that figure in many mythologies, for example, being associated with creativity in indigenous American folklore and associated with death in Norse folklore. Their major competitor for the title of smartest corvid is the New Caledonian crow, which can solve complex sequential problems using different kinds of tools that the crows construct. Parrots are not so advanced as corvids with respect to mechanical problems but can be trained to use language: a grey parrot named Alex could communicate with more than one hundred words.

What They Do

Like other corvids, ravens have keen abilities to see and hear, along with sufficient sense of smell to detect buried food. They can make a variety of sounds and can even imitate human voices as parrots do, including one raven capable of saying the famous word from Edgar Allan Poe's poem, "nevermore." Ravens' perceptual abilities are acute enough that they can recognize specific other ravens and also distinguish between human faces. They can live up to fifty years in captivity. Like most birds and unlike most mammals, ravens form pair bonds for life, lasting up to forty years in captivity.

Ravens solve various problems in their daily lives, including finding a variety of foods, avoiding predators, and finding mates. They can also solve planning problems, such as choosing tools for future use, that go beyond the capabilities of most mammals, including monkeys.

Corvids such as jays, crows, and ravens display complex caching behaviors where they store food for future consumption. They can be deceptive by moving food from caches that have been observed by other birds. Careful experiments have shown that ravens can make decisions about events in the near future (fifteen minutes), as well as in the longer term (up to seventeen hours). Moreover, ravens seem to exert self-control when considering what to do in the future. These findings came from a setup where the ravens put a stone through a tube to cause a platform to collapse and make available a desirable food reward, and also from a setup where ravens exchanged a blue plastic bottle cap for the same reward. Crows have been filmed using sequences of tools to solve problems: pulling up the string to get a small stick that is used to get a big stick that is used to pull out some food.

Like other birds, ravens are capable of associative learning, for example, pairing some human faces with rewards and other faces with punishments.

Moreover, they exhibit reinforcement learning, for example, when they are trained to utter humanlike sounds when they receive food rewards. However, what marks ravens and some other corvids as champion learners is their capability of insight learning, in which causal understanding enables them to come up with new solutions to problems almost immediately. Contrast deep learning in AI, which requires thousands or millions of examples and acquires no causal insights.

A striking illustration of insight learning is a clever experiment where ravens figured out how to get food by pulling on a string to which meat was attached. Because the meat was frozen, they could not rip it off the string, and pecking at the string did not work either. After trying several methods, most of the ravens got the meat by perching above the string and the meat, reaching below the perch, grabbing the string in their bills, pulling the string up, laying the loop of string onto the perch, placing a foot on the string, and bringing the meat up to the perch. The ravens accomplished this complex process without having observed string pulling or being trained to do it. Like people, ravens exhibit individual differences in learning ability, as some of them catch on much more quickly than others.

Such experiments suggest that ravens are capable of causal inference expressed in sensory-motor representations rather than words. The nonverbal inference goes something like "If I pull on the string, then the meat will come closer." We see even more abstract inference in ravens' reaction to the threat of having their meat stolen by other ravens: they are more careful about hiding meat when other ravens are around to steal it. A clever experiment showed that the change in caching behavior is not caused merely by the fact that another raven is looking right at the meat. Caching ravens behave differently depending on whether other ravens in a box have a peephole through which they might be looking at them.

So ravens are capable of complex cognition, but do they have feelings? Bernd Heinrich is convinced that they do: "By a combination of voice, patterns of feather erection, and body posture, ravens communicate so clearly that an experienced observer can identify anger, affection, hunger, curiosity, playfulness, fright, boldness, and (rarely) depression." He remarks that ravens are highly fearful of new situations, a good strategy, given that they spend time around carnivores such as wolves. Heinrich also suggests that ravens have emotional attachment to their long-term mates. Crows sometimes gather around a dying crow, although it is difficult to say whether this is mourning or just curiosity.

Ravens communicate with both their voices and their bodies, but the sounds they utter serve largely to draw attention to themselves, not to convey specific meanings. Although ravens are capable of complicated sounds, their basic sound is just "haw haw," which serves merely to get the attention of other ravens.

How They Work

It is puzzling why ravens and other corvids are capable of such sophisticated behaviors when their brains are much smaller than mammals of comparable sophistication. The two billion neurons in a raven brain are huge for a bird or reptile, but considerably smaller than the more than twenty billion neurons found in great apes such as chimpanzees and orangutans. There is much more to intelligence than brain size measured in either volume or numbers of neurons, as elephants and sperm whales have brains that are much larger in both size and number than humans.

The mental sophistication of ravens is also puzzling because bird brains are so different from mammal brains. In mammals, mental capabilities are strongly correlated with the size of the neocortex, the newer part of the brain that is widely viewed as being responsible for higher kinds of thinking such as planning and decision making. Birds nullify this correlation because they have no neocortex. Bird brains do have some areas in common with mammals, such as the hippocampus, but the overall organization is different. The area of the bird brain that seems to be analogous to the neocortex with respect to carrying out higher cognitive functions is called the nidopallium. Bird brains have the intricate neural circuitry found in mammal brains with connections between the nidopallium and sensory-motor areas.

Because birds fly, they cannot carry around large brains, but they pack in more neurons through smaller neurons and tighter connections. The development of powerful brains that use different means to accomplish similar functions is convergent evolution, just as the octopus eye evolved to accomplish perception differently from the mammalian eye. Compared to reptiles, the larger brains of birds and mammals support areas that integrate sensory representations of different kinds, such as vision and smell.

Despite different organization, bird brains use neural mechanisms for neural firing and synapse modification to support the mental mechanisms found in humans and other mammals. Ravens and other birds have limited language, so we cannot explain their complex behaviors as based on linguistic reasoning. But their visual, auditory, and motor capabilities are substantial, so it is plausible that they carry out their complex inferences concerning tools and caching by using imagistic representations of the world. For example, in figuring out how to pull up a string to get meat, the ravens may use visual images of the string and the meat along with motor images of pulling and holding down the string. Working with such images requires consciousness not just of perceiving something but also of imagining how it could be different. I have not found any experiments testing for imagery in ravens, but evidence that pigeons use imagery comes from experiments similar to those used as evidence for the operation of imagery in humans.

Because of ravens' abilities to discriminate between different kinds of objects and to make generalizations about what to do with them, it is reasonable to attribute concepts and rules to ravens and other birds. For example, ravens can learn that peanuts are good to eat, suggesting that they have a nonverbal concept of peanuts and general rules about eating them. Some researchers claim that crows are capable of analogical thinking concerning the construction of tools, but in chapter 6 I describe flaws in this attribution.

In humans and other mammals, emotions work by a combination of physiological perceptions and cognitive appraisals, using brain areas such as the amygdala, nucleus accumbens, and prefrontal cortex. Birds have a nucleus accumbens and dopamine circuitry for reward learning, but they only have rough analogues of the amygdala and other brain areas important for human emotions. Nevertheless, emotional behavior such as fear suggests that crows use these analogous areas to carry out something like the perceptions and appraisals that support human emotion. Birds are similar to mammals with respect to pain physiology such as nociceptors and pain behavior such as withdrawal, so it is plausible that birds also have conscious experience of pain.

I am not aware of any attempts to determine whether ravens are self-conscious by getting them to recognize themselves in mirrors. But magpies marked with colored dots do behave differently in front of mirrors, leading researchers to attribute self-consciousness to at least one kind of corvid.

Feature Benchmarks

Bees and octopuses each missed several feature benchmarks of intelligence, but ravens score well. They are capable of many kinds of perceiving and problem solving, including planning and deciding about how to get food and avoid competitors and predators. They are particularly impressive with respect to learning, since they are capable of causal insight learning based on limited experience. Ravens are occasionally creative to some extent, as when figuring out how to get meat off a string produces a solution that is new, surprising, and valuable for them.

Some crows can solve problems using twigs to get food and even bend the twigs into the right shape for retrieving objects. They can put rocks into a tube of water to displace the water to make desired food come within reach. They can recognize and distinguish different people and communicate dislike of particular people to other crows. Some of the problems that New Caledonian crows solve require up to eight steps. Crows sometimes exhibit the ability to delay gratification by waiting to eat until a better snack is offered.

Other birds such as pigeons that have been tested more extensively on discrimination tasks have shown themselves to be capable of abstract judgments such as same and different. Ravens are capable of nonverbal inferences, but not verbal reasoning. Pigeons (and wasps) can do transitive inferences of the form "*A* is better than *B*, *B* is better than *C*, so *A* is better than *C*." I would prefer to call this inference rather than reasoning, since birds do not have enough language to give reasons. Similarly, although it is plausible that crows have causal understanding concerning difficult tasks such as manipulating twigs and pulling strings, they lack understanding in the sense of being able to provide explanations.

Ravens do diverse kinds of acting, including aerial feats such as linking talons in midair and playfully rolling down snowbanks. Many of their actions seem tied to feelings, such as responding to pain and emotions such as fear. Ravens perform at least some communicating through limited voice signals and various body movements.

Mechanism Benchmarks

Ravens do much better than bees and octopuses with respect to the mental mechanisms that support human intelligence. Being able to solve insight problems about finding and caching food suggests substantial capabilities of thinking with images. Other birds such as pigeons have been more extensively studied with respect to the operation of concepts and rules, but the behavior of the even smarter ravens suggests that attributions of these kinds of mental representation can legitimately carry over to ravens.

Ravens can make similarity judgments about relations, but not full analogical inference.

Emotions and consciousness probably operate in the brains of ravens, even though these are organized differently from the brains of mammals. The ability of ravens and New Caledonian crows to plan out sequences of actions suggests the operation of the mechanism of intentional action, where actions are represented by motor images.

The major gap in the intelligence of ravens and other birds is clearly with respect to language. The various sounds that ravens make and the complicated productions of songbirds do not add up to language, as they lack the integration of syntax, semantics, and pragmatics described in chapter 2. Parrots such as the famous Alex can acquire a much larger vocabulary and make sophisticated judgments of similarity and difference, but their productions are limited in size and complexity, like the chimpanzees discussed later.

Evaluation

Overall, ravens and other corvids are much smarter than bees or octopuses but still fall well short of human intelligence. With respect to feature benchmarks, ravens score well on perceiving, problem solving, learning, planning, deciding, feeling, and acting. Ravens also get points for rudimentary kinds of abstracting, understanding, creating, and communicating. So ravens pass all the tests for observed intelligence, although at degrees lower than humans.

Ravens also rate well with respect to mechanisms of imagery, concepts, rules, emotions, intentional action, and consciousness, even though their brains are organized very differently from mammals. On the other hand, they lack basic mechanisms for language and are challengeable with respect to analogy. Among my selected animals, ravens surpass bees, octopuses, and dogs and seem the equal of dolphins and chimpanzees. However, by virtue of powerful language abilities and broader versions of problem solving and learning, humans far surpass ravens and other birds.

Dogs

The animals most familiar to people are cats and dogs, with more than a hundred million of each owned as domestic pets worldwide. People get attached to their pets and attribute complex thinking to them, such as emotions that include love and guilt. Much more research has been conducted on dogs than on cats, giving rise to the systematic study of "dognition." How smart are dogs compared with humans and other animals?

Dogs became part of human society more than ten thousand years ago when wolves became accustomed to hanging around human encampments. Technically, dogs are still part of the original species of gray wolves from which they evolved, because dogs and wolves can produce fertile offspring. But evolution and breeding have produced differences in dogs with respect to both appearance and behavior. For example, wolves mate for life and produce only one litter per year, whereas dogs are promiscuous and can produce up to two litters per year. Originally dogs were bred for particular purposes such as hunting, but since the nineteenth century, breeds have been produced for aesthetic reasons.

What They Do

Dogs are capable of actions that include walking, running, jumping, eating, and making sounds. Their ability to operate in the world depends on acute senses, including a sense of smell that is more powerful than in humans. Dog brains have ten times the olfactory neurons of people, enabling greater effectiveness at detecting faint smells and making fine discriminations. Dogs also have acute hearing, but their visual abilities are weaker than those of humans, and dogs are less able to notice fine details and make color discriminations. Using vision and smell, dogs can recognize and distinguish familiar people.

Dogs solve a wide variety of problems, including complex feeding behaviors such as hunting and searching for food. Dogs are also capable of solving navigation problems such as getting from one yard to another and moving around detours. Like bees, dogs are capable of finding shortcuts to desired locations, suggesting that dogs have cognitive maps of their environment rather than just going from landmark to landmark. Most pet dogs are neutered, but feral dogs are capable of finding other dogs to mate with. These kinds of problem solving require some planning and decision making, for example, choosing which direction to go to get around an obstacle.

All problems involve goals that provide motivation, and dogs have an unusually strong motivation to interact with people resulting from the selection that has occurred over thousands of years. People have chosen dogs who get along well with them. A fascinating experiment in Russia over the past fifty years found that foxes can also be domesticated to become much more friendly. Because dogs have goals to interact with people, they have a diverse set of behaviors for encouraging interactions such as making eye contact.

Like ravens, dogs perform several kinds of learning. They can remember locations where desired objects are stored. They engage in associative learning, as in Pavlov's famous experiment in which dogs learn to associate food with the ringing of a bell. Dog training relies heavily on reinforcement learning, where dogs are given food as a reward for desired behaviors such as performing tricks. Like people, some dogs are faster to learn than others, and old dogs are usually slower to learn.

Dogs are also capable of social learning, where they acquire behavior by imitating other dogs or people. For example, a dog can learn to detour around a barrier by watching another dog or watching a human. Such learning happens much faster than the many trials required for associative and reinforcement learning. Dogs can learn by imitating the behavior of other dogs, but teaching has only been established in a few species, such as meerkats. Like many animals, dogs enjoy playing, which can contribute to learning motor skills, hunting techniques, and social rules.

Are dogs creative in the sense of coming up with problem solutions that are new, surprising, and valuable? They are capable of learning to pull on strings to get food attached to the end of the string, which would seem to count as new and valuable. However, dogs do not have a deep understanding of what they are doing, because they also pull on strings that are not connected to food. Dogs are capable of some abstraction, for example, learning how to classify pictures of dogs as opposed to ones of landscapes.

People who are emotionally attached to their pets and treat them as family members are naturally inclined to assume that dogs have feelings much as people do. To avoid both anthropomorphism (dogs are just like us) and anthropocentrism (dogs are totally different from us), the best strategy is to look carefully at their behavior and brain processes.

We have two good reasons for concluding that dogs experience pain. First, after injuries such as a damaged paw, dogs show some of the same behaviors that occur in humans. Injured dogs become more vocal, with greater degrees of yelping, whining, and howling, and they may show more general changes, such as trembling or agitation. A dog with a damaged paw may limp and attempt to groom the paw by licking it. Second, the physiological mechanisms of pain are similar in all mammals. Dogs have nociceptors in their skin that send signals to brain areas similar to those of humans such as the amygdala. The evidence that dogs feel pain is much stronger than the evidence that bees or octopuses feel pain on both behavioral and physiological grounds.

The perceptual experiences of dogs with different smells, sounds, and sights are important for their ability to solve problems and learn. Behavior and physiology both support the view that dogs have smells, sounds, sights, taste, and touches, all of which furnish people with conscious experiences. Dogs also seem to be able to have multimodal integration, for example, recognizing a person by sight, sound, and smell. So perception in dogs plausibly comes with conscious feelings.

Similar arguments support the claim that dogs feel at least simple emotions such as happiness, sadness, fear, anger, and aggression. There are behavioral indications such as the intense activity of a dog that is excited about going outside or the growling and cowering of a dog that is afraid of a larger dog. Moreover, the brain areas known to be important for human emotions, such as the amygdala and the orbitofrontal cortex, are also found in dogs and other mammals. Therefore behavior and physiology converge to support attributing emotions and their attendant feelings to dogs. Dogs can discriminate emotional expressions on human faces. It has recently become feasible to use brain scanning to determine that dog emotions show activation in roughly the same brain areas as in humans.

Sometimes such attribution goes too far by ignoring alternative explanations. People naturally think that their dogs are feeling guilty when they put their paws over their heads after some transgression such as making a mess. But careful experiments have found that such behavior is better understood as merely an attempt to avoid punishment, because it occurs whether or not the dog has actually done anything. Humans threatened with punishment for something they did not do would more likely display outrage than cowering. On the other hand, the Attribution Procedure supports the view that dogs can experience jealousy, as shown in chapter 6. Theorizing in the study of animal behavior is shifting away from simplistic ideas about dominance hierarchies toward the recognition of much more complex social networks based on emotional bonds of attachment and affiliation.

Higher feelings in humans include self-consciousness, but dogs have not been found to recognize themselves in mirrors when using the test of applying a mark to them that they can recognize. However, the sense of smell is much stronger in dogs than the sense of sight, so it might be that dogs have self-consciousness based on how they smell to themselves. Dogs at least can tell the difference between the smell of their own urine and the smell of the urine of other dogs.

We would be rash to suppose that dogs have exactly the same feelings as humans with respect to complex emotions such as guilt and abstract representations such as self-consciousness. But feeling seems to be a substantial part of the cognition and intelligence of dogs.

Dogs are capable of communicating with people and with other dogs. They can send nonverbal signals via body parts that include the snout, mouth, eyes, ears, and tail, which can carry different information depending on spatial positions and form and intensity of movement. Dogs also use pheromones to send chemical signals by smell. Dogs have a repertoire of auditory signals that include growls, whines, whimpers, howls, and barks that vary in harmonics and frequency. Dogs are also capable of understanding communications from people who do not use any of these signals. Dogs can follow the pointing and even the gaze of people, and many dogs can follow verbal commands.

How They Work

Dog brains have approximately the same number of neurons as raven brains, around 2.2 billion. This similarity is surprising, given that dog brains are much larger than bird brains, which are more densely connected. Cats are usually much smaller than dogs and have about half as many neurons, although with slightly greater neuronal density. Unlike the brains of birds, the brains of dogs and other mammals have similar structures to the brains of humans. The big difference is that humans have far more neurons (eighty-six billion), including far greater numbers in the neocortex. Dog neurons are similar to human neurons, with the same chemical processes based on neurotransmitters and hormones. As in humans, the neurons in dog brains can represent the world by means of patterns of firing in groups of neurons.

Perception in dogs works much the way it works in humans, through stimulation of receptors for the various senses followed by interpretation carried out in the brain. Humans are able to turn their perceptions into images that can be manipulated in various ways, for example, by rotating visual images and combining olfactory images, as when you imagine how to combine the taste and texture of maple syrup and peanut butter. Although evidence suggests that humans have olfactory images, I have been unable to find any experiments on imaging using smell in dogs, even though their sense of smell is more acute and important than for people.

However, experiments do suggest that dogs are capable of visual imagery just as humans are. As with bees and ravens, evidence for imagery includes showing that dogs are capable of navigation shortcuts that take them directly back to where they started rather than laboriously retracing their paths. Dogs make such shortcuts, and they are also capable of learning how to make detours efficiently. The most plausible explanation of shortcuts and detours is that dogs, like rats and other animals, have a cognitive map of their environment that captures aspects of its spatial organization. In humans, consulting a cognitive map, such as your mental map of the city where you live to find the best route to where you work or study, requires forming a mental image of the map and then manipulating it in various ways, such as imagining different routes. Analogously, it is plausible that dogs have visual images of their environment which they manipulate with conscious experiences.

Various kinds of evidence suggest that dogs have concepts, although these lack the verbal richness found in people. Dogs are able to make perceptual discriminations among classes of objects, such as identifying people and drugs. Moreover, highly trained dogs are able to associate verbal commands with hundreds of different objects and actions. Dogs also are capable of operating with rules, such as the association *if dish, then food*. Trained dogs have multimodal rules, such as the nonverbal equivalents of *if hear "sit," then sit*. I have not found any studies that claim that dog cognition employs analogies.

In contrast, abundant evidence shows that dog thinking is heavily influenced by emotional mechanisms. Dog owners naturally think that their pets are happy when given a treat and afraid when confronted by a menacing larger animal. In these cases, dog behavior could alternatively be explained by simpler mechanisms of responses to rewards and threats. But I argue in chapter 6 that dogs and other mammals are capable of grief and jealousy arising from the emotional bonds between pet and owner. The pet may not feel exactly the same love toward the owner that the owner feels toward the pet, but something like affection and attachment is apparent. Dogs even show separation anxiety when their owners leave, just as human babies react negatively to separation from their caretakers.

Brain anatomy explains why dogs can have emotional reactions similar to those found in people. Like other mammalian brains, dog brains have the anatomical areas known to be important for emotions in people, including the amygdala, basal ganglia, and prefrontal cortex. The fact that the prefrontal cortex is much smaller in dogs than in people may mean that dogs are incapable of the most complicated kinds of human emotions such as guilt, schadenfreude (joy in the misfortune of others), and fear of embarrassment.

There has been little investigation of whether dogs perform actions as the result of intentions. In one experiment, dogs appeared to intend to communicate to their owners the location of hidden food by looking frequently at both the owner and the food. An extraordinary dog that was trained to use simple signs for making requests seemed to be intentionally asking to be petted, given drink or food, or taken for a walk.

Dogs are capable of auditory and bodily communication, but does this amount to language? Growls and postures make effective signals, but they lack the syntactic complexity of human language. Highly trained dogs may be able to understand commands such as "get ball" that require a minimum of syntax and semantics, but we have no evidence that dogs have the capability of more complicated kinds of comprehension or generation. At best, dogs only have the beginnings of the linguistic mechanisms that contribute so much to human intelligence.

When a dog gets a thorn in its paw, behavior such as whimpering and grooming the paw suggests that the dog is experiencing pain. Other phenomena that support attributions of consciousness include perceptual operations such as smell, use of images in cognitive maps, and a variety of emotions. Despite a much smaller neocortex than humans, dogs appear to have conscious experiences in the same way that people do through the formation of neural representations that can be combined into more complex representations and then compete against other representations to become conscious. However, dogs may not have the full range of consciousness that people do if dogs lack the robust sense of self that operates in humans.

Feature Benchmarks

On benchmarks for features of human intelligence, dogs perform well, with substantial degrees of perceiving, problem solving, learning, feeling, acting, and communicating. More limited are capabilities for abstracting, understanding, and creating. The problem-solving activities of dogs seem to indicate some capacity for planning and deciding, but not with the complexity of timing and options found in humans.

Mechanism Benchmarks

The mechanisms of intelligence in dogs are similar to the mechanisms that operate in humans through similarly structured but much smaller brains. The ability of dogs to accomplish problem solving, learning, and other features of intelligence derives from mechanisms for imagery, concepts, rules, emotions, and consciousness. Dogs may also be capable of basing actions on intentions, but we have no evidence that dogs are capable of using analogies. Similarly, despite their communicative abilities, dogs seem to have little of the linguistic capacity of humans. Among mammals, the most impressive language-like ability is found in prairie dogs, which are not dogs at all; they are rodents capable of making sounds that communicate to other prairie dogs information as complex as the size and clothing color of humans.

Evaluation

Overall, pet owners are right to be impressed by the intelligence that their dogs show in problem solving and learning. Dogs are remarkable in their ability to interact with people and to become emotionally attuned with them. This ability results from more than ten thousand years of evolutionary selection for dogs that are best suited to get along with people. In raw intelligence, dogs may not quite measure up to the corvids that can do more complicated kinds of multistep problem solving.

Breeders have speculated about individual differences between dogs and between breeds of dogs. However, claims such as that border collies are the smartest breed and Afghan hounds are the dumbest are based on unsystematic observations rather than controlled experiments. Nevertheless, dogs display individual differences with respect to their speed of learning and capacity for problem solving, and I hope future research will provide neural explanations for these differences.

Dolphins

Since the 1970s, scientists have widely regarded dolphins as smart and sociable, with powerful brains and sophisticated language. Dolphin intelligence is impressive, but no more so than that of corvids and the great apes. Dolphins are cetaceans, the class of animals that include whales and porpoises. Dozens of different species of dolphins exist, but most research has been done on a particular species, the bottlenose dolphin.

Cetaceans evolved around fifty million years ago from land mammals akin to the modern hippopotamuses that spent more and more time in the water. Eventually they became well adapted to full-time aquatic existence, and some of them, such as sperm whales and dolphins, evolved large brains that support complex problem solving. Dolphins, apes, and corvids exemplify convergent evolution, where different evolutionary paths led to the same result: brains that support a high degree of intelligence.

What They Do

Because they live underwater most of the time, dolphins have no sense of smell and minimal sense of taste. But their eyesight is good both in and out of the water, and their hearing ability is much better than that of humans and dogs, including being able to detect high-frequency sound. This sense is a major contributor to their navigational abilities because, like bats, dolphins can use echolocation to detect objects far away. Dolphins send out clicks that bounce off distant objects and are reflected back to the animal, which can use them for identifying objects by their size and density. One species of dolphin also has the rare capacity to sense prey by identifying their electric fields.

Observations in the wild and laboratory experiments show that dolphins are capable of complex problem solving. Bottlenose dolphins use air bubbles to capture prey, and some dolphins use sea sponges on their heads to help access food. A pair of dolphins at Disney's Epcot Center learned to solve problems that required retrieving weights by observing human divers and came up with their own efficient way of collecting weights. In contrast to the usual reputation of dolphins as gentle and kind, male dolphins cooperate to herd female dolphins for mating purposes.

Dolphins are also able to solve communication problems by whistles and pointing. Many of these problems require planning, at least over short periods of time, and some of them require deciding, for example, between two buckets of fish. To solve problems, dolphins are capable of actions that include swimming, attacking, and mating. Males sometimes kill the young offspring of other dolphins, as well as young porpoises.

The learning capabilities of dolphins go well beyond merely making associations between perceived stimuli. They are able to comprehend a wide range of concrete and abstract concepts, including same or different shapes, fewer or more dots, larger or smaller objects, humans versus dolphins, and descending or ascending tones. Dolphins are adept at imitating the problem-solving strategies of humans and other dolphins, and a group near Adelaide learned to walk on their tails by imitating a dolphin that had learned the trick by imitating trained dolphins.

Anecdotal evidence suggests that some dolphin mothers teach hunting techniques to their offspring. A few laboratory experiments, such as the one with weights, suggest that dolphins are capable of learning by insight, quickly grasping a solution rather than relying on trial and error. Some researchers claim that dolphins are not only capable of creatively coming up with new and useful problem solutions but also can be trained to produce a higher rate of novel solutions.

Evidence that dolphins are capable of understanding is limited but suggestive. A dolphin in Florida was trained to discriminate between high and low tones and also to recognize when it was hard to tell between these two classes. One interpretation is that the dolphin was capable of metacognition that recognizes uncertainty in its own thinking. Some researchers have suggested that dolphins are capable of attributing mental states to other dolphins, for example, while following their gaze.

With their smiley faces and energetic actions, dolphins appear to have feelings, but we must always consider alternative explanations. Dolphins appear to smile not because they are happy but simply because their mouths are naturally curved. On the other hand, reports of several species of dolphins say that they display nurturant behavior toward dead calves by carrying the young. Dolphins exhibit emotional contagion, adjusting their behaviors based on those of other dolphins, but whether this amounts to empathy is unclear. Cetaceans have nociceptors and pain circuitry similar to humans, and unlike fish, dolphins have complex, mammalian brains, so there is little doubt that they feel pain. Moreover, in careful experiments, dolphins have been shown to recognize themselves in mirrors, suggesting that they have self-consciousness beyond the minimal awareness of pain. I discuss the ethical significance of such pain and suffering in chapter 7.

Dolphins communicate by means of auditory signals such as whistles and bodily signals such as pointing. Whistles transmit information about what a dolphin is currently doing, such as foraging and socializing. Individual dolphins have signature whistles that convey their identity to their offspring and other dolphins. Whistles indicate the direction of travel and the presence of food. Dolphins are intensely social, and auditory communication helps to provide coordination, but early claims that their whistles constitute a language have not held up to critical scrutiny.

How They Work

Dolphin brains are much bigger than those of dogs, and some species of dolphins have larger brains than humans. For bottlenose dolphins, brains weigh around 1.8 kilograms, or 4 pounds, slightly larger than human brains. Some species of dolphins that for historical reasons are known as whales, the killer whale (orca) and pilot whale, have more neocortical neurons than humans. As already mentioned, the relationship between brain size and intelligence is only approximate.

Because dolphins are mammals, their brains have roughly the same structure as those of dogs and people, but there are subtle differences. Compared to other mammals, dolphins have smaller frontal lobes and hippocampus, but larger auditory areas. Dolphin brains also have a special kind of neuron called spindle cells that are thought to contribute to the cognitive flexibility of apes and humans.

Dolphins' ability to make discriminations among objects and situations suggests that their neural mechanisms support concepts, such as *sponge*. Similarly, through association, reinforcement, and insight, dolphins construct rules, such as that if they leap into the air in front of a human trainer, then they will get fish.

I have not been able to find any studies that say whether dolphins think with mental images. In humans and other animals, the hippocampus is important for memory and cognitive maps, so perhaps imagery in dolphins is limited by their comparatively small hippocampus. On the other hand, because of their powers of echolocation, the natural sort of imagery to look for in dolphins would be auditory, but such tests are hard to conduct. Another possibility would be to investigate whether dolphin navigation takes shortcuts that are naturally explained in terms of visual imagery using cognitive maps. Also missing in research on dolphin cognition are studies that address whether dolphins are capable of thinking by analogy.

Whether dolphins have intentional actions is likewise difficult to determine. They do have representations of actions, as they can be trained to repeat the previous action, but intentional action also requires combining the representation of the action with wanting to do the action. Intentional action appears to occur in complex dolphin behaviors such as working together to herd small fish up to the surface where they can be easily eaten.

Dolphins communicate information using whistles, pulses, and body language, but the result falls far short of human language. The dolphin expert Justin Gregg evaluated dolphin language with respect to ten ingredients of human language:

1. Limitless expression of ideas or concepts
2. Discrete combinatorial system that combines small units into meaningful combinations
3. Recursion that embeds syntactic structures inside other structures
4. Special memory for words or symbols
5. Displacement across time to represent events in the past or future
6. Use of environmental input to learn by interaction with skilled speakers
7. Arbitrary relationship between symbols and what they describe
8. Going beyond emotion so that linguistic concepts are not just about internal emotional states
9. Generation of novelty in learning and creating new ideas
10. An aptitude for social cognition

Gregg allowed that dolphin communication has some of these ingredients, such as the arbitrary relation between whistles and what they convey. But he rated dolphins badly on other ingredients, such as limitless expression and having a full combinatorial system. We should be impressed with the communication abilities of dolphins in the intricacy of their whistles without supposing that they approximate human language.

Gregg judged that dolphins have simple emotions such as pain, pleasure, fear, anger, and sadness, along with sensations such as hunger, thirst, and sexual desire. His conclusion was partly based on the fact that all vertebrates possess the same subcortical regions like the basal ganglia, known to be responsible for mediating arousal states. But Gregg was skeptical about whether these emotions carry with them subjective experiences and awareness of emotions such as fear. This skepticism makes it unclear whether what he attributes to dolphins is fear or just some more primitive way of detecting and responding to threats. In chapter 6, I argue that it is reasonable to attribute emotional consciousness to dolphins and other mammals.

Dolphins feel pain and other sensations, and fear and other emotions, through neural mechanisms for consciousness that include representation of the world and competition among representations. Moreover, the experiments that found dolphins capable of recognizing themselves in mirrors indicate self-consciousness. This conclusion can be challenged because of difficulties in interpreting the mirror test and in determining the nature of the dolphins' self-representation. But preponderance of evidence in the Attribution Procedure supports the conclusion that dolphin cognition employs the mechanism of consciousness.

Feature Benchmarks

On features, dolphins score well with respect to perceiving, problem solving, learning, feeling, communicating, and acting. The evidence is weaker but still credible that they have elements of understanding, planning, deciding, abstracting, and creating. The degree of accomplishment of these features falls short of what humans can do, for example, with respect to the range of problems solved and the types of learning. Dolphins are only good at the small range of problems in their environments, and their learning does not extend to hidden causes.

Mechanism Benchmarks

Dolphin brains have the same basic neural mechanisms as humans, and their brains are around the same size, so why dolphins are not as smart as people is puzzling. The big difference concerns language, where humans have full combinatorial capacity for syntax and semantics. Language gives people an enormous advantage in social groups, where linguistic communication contributes to cooperation and social problem solving and also in the transmission of problem solutions across generations. Such transmission became all the more powerful once writing was developed less than ten thousand years ago. Another big advantage for humans is that the flexibility of our hands enables us to use a wide variety of tools in problem solving, whereas the best that dolphins have been able to do with tools is to balance sponges on their heads.

The syntactic and semantic richness of language has probably also given humans the capacity to think with hidden causes and analogies. Further study may provide evidence about whether dolphin cognition operates with imagery, particularly auditory imagery. On the positive side, the mental mechanisms of dolphins include concepts, rules, emotions, intentional actions, and consciousness.

Evaluation

The large brains and behavioral repertoire of dolphins imply that they are in the same ballpark as apes and corvids with respect to intelligence, although claims that dolphins are cognitive rivals of humans have not stood up to investigation in recent decades. There is no one-dimensional test of intelligence across species, and it is meaningless to estimate dolphin IQ when dolphins are incapable of taking human IQ tests. Dolphins display kinds of intelligence that are not measured by IQ tests, such as kinesthetic, emotional, spatial, and social abilities. Another way to be impressed by dolphin intelligence is to consider how much better they are than current AI systems on some of the feature and mechanism benchmarks.

Chimpanzees

Humans and chimpanzees diverged from a common ancestor around seven million years ago. Bonobos were once thought to be just a version of chimpanzees but are now recognized as a separate species that diverged from chimpanzees around one or two million years ago. Chimpanzees and bonobos are genetically similar but have important differences. Bonobos engage in recreational sex, rely on mothers until adulthood, share with strangers, and form female alliances based on frequent sexual engagements. Chimpanzees use tools in the wild, but bonobos are more adept at peaceful social interactions, although they can use tools when these are made available in captivity.

Chimpanzees and bonobos inhabit separate environments divided by the Congo River. Researchers have offered a variety of explanations about how bonobos evolved to have societies organized more by female alliances than by male dominance hierarchies. The native habitat for bonobos consists of forest where fruit is readily available, whereas chimpanzees inhabit a wider variety of habitats where getting food can be more challenging. The move away from aggressiveness in bonobo society had an advantage in eliminating infanticide, which is a common feature of chimpanzee and dolphin societies, where males kill the offspring of other males.

What They Do

Chimpanzee sensory systems are similar to those of humans. Most mammals, including dogs and dolphins, have limited color vision because they only have receptors in their eyes for two basic colors. Like other apes, monkeys, and humans, chimpanzees have receptors for three colors. Birds, including ravens, have even more sensitive color vision thanks to receptors for four colors that allow sensitivity to ultraviolet light.

Chimpanzees in the wild solve problems that include finding food, mating, avoiding predators, getting along with the approximately one hundred other members of their community, and battling with other groups. In dealing with the physical world, chimpanzees accomplish kinds of acting such as moving and eating. They are also skilled at using tools to get food and water, working with modified sticks to obtain termites and ants, stones to crack nuts, and leaves to sponge up water. The modification of sticks and the compression of leaves show that chimpanzees are capable of making tools for future use, but I have not seen any reports of them making tools that can be used to make other tools.

Chimpanzees engage in social problem solving in various contexts, including grooming to remove parasites, hunting together to catch monkeys to eat, warring against other communities, and males cooperating to control females for mating. In contrast to chimpanzee battles against other groups, when dispersed bonobo communities encounter each other, the typical interaction is sexual contact among females rather than fighting among males.

Effective tool use by chimpanzees displays causal understanding of the physical world, for example, when striking a nut with a rock causes the nut to crack. Bonobos have a rich understanding of their social worlds, for example, when females cooperate to help other females. Scientists debate the extent to which chimpanzees understand each other's minds. Evidence suggests that they can at least make inferences about what other chimpanzees are looking at, but attempts to show that chimpanzees infer each other's beliefs have been inconclusive. Like some corvids, chimpanzees engage in acts of deception that require making inferences about what other chimpanzees can be expected to do, but inferences about behavior are easier than inferences about another's mental states, which are nonobservable causes. Some experiments suggest that chimpanzees are capable of using analogies, but I challenge the validity of this interpretation in chapter 6.

Deception and cooperation show that chimpanzees are capable of planning, which also occurs in other domains. For example, chimpanzees have been observed saving up piles of stones or feces to throw at humans. Chimpanzees are capable of making decisions that exhibit self-control by delaying a reward to get a greater reward, just as in the famous marshmallow task used with humans, where children are offered a choice between one marshmallow now or two marshmallows later.

The use of tools among chimpanzees varies in different communities, so it is acquired by learning and cultural transmission rather than innate neural wiring. Some chimpanzees learn the effectiveness of sticks for obtaining food by trial and error, but useful techniques spread through communities by imitation learning. One chimpanzee can watch another using a tool effectively and quickly pick up the same skill. Like other animals, chimpanzees can slowly learn by association and reinforcement, but they can also learn rapidly by insights into the causal structure of their environments. Insight learning was first documented by Wolfgang Köhler in the early twentieth century when he observed chimpanzees who figured out how to move boxes to obtain bananas. Although imitation learning is common in chimpanzees and bonobos, teaching in which one individual intentionally gets another to learn is rare.

Chimpanzees can learn some abstract relational concepts such as identity and difference, can count up to seven, and can sum small sets of objects. The insight learning of new problem solutions amounts to a kind of creativity, as did the original invention of the various tools used by chimpanzees.

Because chimpanzee physiology and brain structure are so similar to those of humans, many of their behaviors are plausibly explained by supposing that the animals have similar feelings to us. The strongest case is for relatively simple feelings such as pain, happiness, fear, anger, and surprise. Chimpanzees seem to have recognizably different personalities, with characteristics such as conscientiousness and openness to experience predicting their ability to solve problems. Researchers have also attributed to chimpanzees much more complicated emotions such as consolation, altruism, and sensitivity to unfairness.

Chimpanzees communicate with one another using vocal calls that include pant-hoots, whimpers, screams, squeaks, barks, coughs, grants, and a form of laughter. Chimpanzees also communicate nonvocally with facial expressions, body postures, and dozens of different gestures, such as stroking to request food and arm raising to request grooming.

How They Work

Chimpanzee brains are about one-third the size of ours, with approximately twenty-eight billion neurons. Chimpanzee and bonobo brains are similar in size and structure, although they have subtle differences. Brain scans found that bonobos have more gray matter (neurons, rather than the white matter consisting of connections between neurons) than chimpanzees in the hypothalamus and parts of the amygdala and interior insula. These differences may partly explain why bonobos are less aggressive and more sexual than chimpanzees. Chimpanzee neurons are organized into the same brain areas as in humans, although with a smaller prefrontal cortex. Aside from sheer size, no characteristics of chimpanzee brains explain why people are generally smarter.

Three kinds of evidence support the hypothesis that chimpanzee minds use visual mental imagery. First, their navigation through complex environments is highly efficient, suggesting that they are capable of working with spatial maps that provide shortcuts. Second, the occurrence of insight learning in problems such as moving boxes to reach a banana suggests ability to work with mental images. Such inferences could hypothetically be made with words, but chimpanzees do not have many words to work with. Third, monkeys are adept at learning by imitation to perform tasks such as using tools to find food. Learning by imitation is most effectively done by mentally representing the motions of another chimpanzee or human and transferring them into one's own motions.

The antiquated understanding of concepts as essentially verbal implies that only humans have concepts, but experiments and observations support the conclusion that chimpanzees and other primates work with an abundance of concepts. Chimpanzees can make perceptual discriminations such as telling different kinds of fruit apart. They are also capable of working with more abstract concepts including sameness, difference, number, and summation. Concepts provide mechanistic explanations of how minds can organize objects into useful categories, as chimpanzees frequently accomplish.

Chimpanzees are also capable of combining concepts into rules such as *If you eat a banana, then it tastes good*. Such rules are multimodal in that concepts such as eating, banana, and tasting good have sensory-motor rather than verbal representation in chimpanzee neural groups. A behaviorist would say that these are not really mentally represented rules, just simple associations between stimuli and responses. But chimpanzees are capable of operating with more complex rules such as *If you put a nut on a big rock and hit it with another rock, then the nut cracks and you can eat the inside.* Insight learning requires chaining together a sequence of rules, such as the following combination: *If you move boxes, then you can stand on them; and if you stand on the boxes, then you can reach the banana; if you can reach the banana, then you can grab it and eat it.* So rules understood as if-then mental representations contribute to chimpanzee thinking.

Do chimpanzees have intentions that help to explain their actions? If they solved problems merely by trial and error, then we would have no reason to attribute intentions to them. But researchers have studied many cases of more deliberate problem solving with long-term planning, cooperation, and insight learning that are plausibly explained by supposing that chimpanzees' actions are intentional.

Moreover, experiments suggest that chimpanzees have some understanding of the intentions of other chimpanzees and humans. They treat the actions of humans as goal-directed, and they are able to complete the failed actions of others. They provide specific help to partners based on an understanding of their partner's needs, and they discriminate intentional actions and accidental ones. Experiments on chimpanzee self-control suggest that they are capable of metacognition, thinking about their own state of thinking. They engage in delay of gratification by distracting themselves. Therefore attributing intentions to chimpanzees is reasonable.

Goals are not just cold cognitive representations but are bound with emotional states such as desire and craving that provide the motivation to accomplish the goal. Many humanlike emotions have been observed in chimpanzees, including fear, anger, excitement, surprise, and grief. Chimpanzee brains have all the emotion-relevant brain areas that operate in humans, although the prefrontal areas required for cognitive appraisal are not as developed. Nevertheless chimpanzees have sufficient brainpower to appraise the relevance of situations to their basic goals such as feeding, mating, and social status. Moreover, their strong anatomical similarities with humans suggest that chimpanzee brains are getting the same kinds of physiological inputs that contribute to human emotions. With many billions of neurons, chimpanzee brains have no problem binding together representations of situations, appraisals of those situations, and physiological inputs to produce many of the same emotions found in humans.

I suspect that chimpanzees are capable of mixed emotions, such as a combination of curiosity and fear generated by an encounter with an unusual object. But for reasons presented in chapter 5, I doubt that they can experience nested emotions such as fear of embarrassment and regret for anger. Moreover, complicated emotions such as embarrassment, guilt, shame, and gratitude may require more complex social understanding than nonhumans can accomplish. However, the capacities of chimpanzees for consolation and grief show that they are not limited to the simplest emotions.

Because chimpanzees operate with much the same mechanisms of pain, perception, imagery, and emotions as human beings, it would be surprising if they did not also have similar consciousness. Chimpanzee brains have ample numbers of neurons capable of representation, binding, and competition to support conscious experience. Chimpanzees even perform better than humans on some memory tasks, such as memorizing the locations of numbers on computer screens.

In addition, experiments show that chimpanzees recognize themselves in mirrors, implying that they have self-consciousness. The constant concern of chimpanzees and bonobos with their status in social groups suggests that they are also aware of how they stand in relation to others.

The major mental mechanisms missing in chimpanzees concern language. Attempts to teach human language to chimpanzees and bonobos have seen limited results. The most linguistically sophisticated ape of all time was Kanzi, who acquired the ability to understand complex utterances such as "put the key in the refrigerator." But Kanzi's limited ability to generate utterances was below the level of a three-year-old human child. Although chimpanzees have the means of vocal and bodily communication described in the last section, they lack important language mechanisms for embedded syntax and semantics, discussed in chapter 5.

Feature Benchmarks

The ways in which chimpanzees deal effectively with their physical and social worlds show that they possess the twelve features of human intelligence: perceiving, problem solving, planning, deciding, understanding, learning, abstracting, reasoning, creating, feeling, communicating, and acting. While recognizing chimpanzees as intelligent, we should acknowledge that they do not have these features to the same degree as humans.

Chimpanzees can abstract from current situations with concepts such as *same* and *different* but cannot move beyond their sensory experiences as humans can in developing mathematics, science, and philosophy. Perhaps similar limitations impair chimpanzees' abilities to take communication into the full realm of language and take creating beyond the sensory-motor contexts of relatively simple problems. In chapter 5, I provide a unified neural explanation of why people are much better than chimpanzees and other animals at going beyond the information given.

Mechanism Benchmarks

Chimpanzees rate highly on mental mechanisms operating in their large brains. The best explanations of their full range of features imply that chimpanzees are capable of operating with imagery, concepts, rules, emotions, and intentional actions. But they lack full language mechanisms, a lack that also costs them the ability to accomplish the full range of analogical inference (see chap. 6). Chimpanzees have some of the rudimentary mechanisms of language, such as learning word strings, but experts on animal intelligence largely agree that full-blown language is found only in our own species.

Consciousness is always more controversial, and chimpanzees' linguistic limitations may hinder the complexity of the objects of chimpanzee consciousness. Nevertheless, mirror self-recognition and other features of chimpanzee intelligence suggest that they also possess consciousness.

Evaluation

As our evolutionarily closest surviving relative, chimpanzees unsurprisingly rank close to humans on features and mechanisms of intelligence. The main differences are matters of degree with respect to features such as abstracting and a sharper matter of kind with respect to the mechanisms of language. Chimpanzees are superior to all current robots and AI systems with respect to their capabilities for multimodal perception, cooperative problem solving, emotions, and learning by insight and imitation.

Animal Report Cards

I have examined six smart animals representing insects, mollusks, birds, and mammals. All the species rank among the most intelligent in their respective orders, except for dogs, which I have included because of their powerful social connection with humans. Figure 4.1 gives an idea of the evolutionary history of the six species and humans, showing diverse paths to intelligence.

Table 4.1 provides more details of the evolutionary history and brain size of seven species. Besides brain size, which scales with intelligence only approximately, factors important for intelligence include the ratio of brain to body size, the organization of the brain into functional areas, the density and folding of the brain, and the quantity of neocortex or analogous areas in bird brains.

Applying the twenty benchmarks for features and mechanisms to six smart animals shows how impressively intelligent they are, as report cards 4.1 and 4.2 reveal. As in chapter 3, an A means that the animal is close to humans, B means that the animal has much of human capability, C means that the animal has a bit of human capability, and F means a total lack of human capability. Lack of evidence that would enable grading is indicated by a question mark.

Even bees approximate human intelligence in many respects, and chimpanzees come very close. The major gap is the absence of mechanisms of

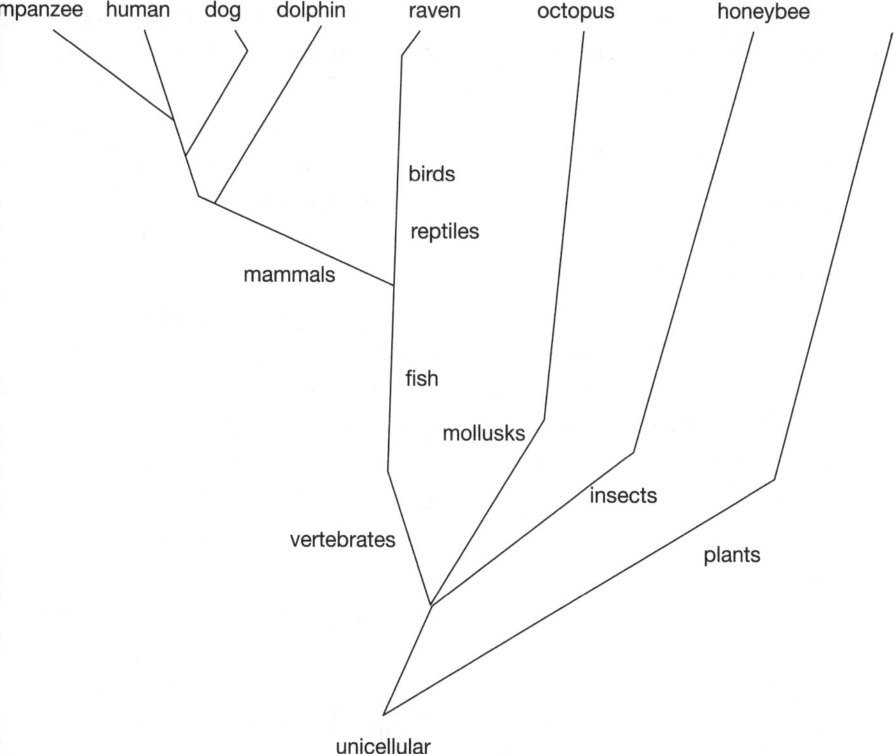

Figure 4.1
Approximate evolutionary history of smart animals, showing divergences. The lengths of the lines are not proportional to time.

Table 4.1
Brains and evolutionary history of seven species

Animal	Evolved years ago	Brain weight	Brain volume	Number of neurons
Bee	125 million	2 mg	1 mm^3	960,000
Octopus	300 million			500 million
Raven	17 million	15 g	16 cm^3	2.1 billion
Dog	15 thousand	70 g	80 cm^3	2.2 billion
Dolphin	15 million	1,500 g	1,500 cm^3	37 billion (cortex only)
Chimpanzee	4 million	420 g	400 cm^3	28 billion
Human	200 thousand	1,300 g	1,260 cm^3	86 billion

All numbers are estimates. Octopus neurons are distributed across their head and arms. Species of dolphin vary greatly in brain size.

language in nonhuman animals, which I analyze more thoroughly in chapter 5. Much room remains for disputing particular grades, but the report cards have a big advantage over monolithic views of intelligence such as IQ and *g*: they indicate that intelligence is not a single factor but is distributed over at least twenty features and mechanisms. Each of these characteristics is a matter of degree, rather than a binary matter of having it or not.

The report cards for animals also provide comparisons with machine intelligence displayed in report cards 3.1 and 3.2. The comparison shows that animals generally have more features and mechanisms of intelligence than most machines, although some artificial intelligence programs, such as IBM Watson, do better on linguistic dimensions. Moreover, we can expect artificial intelligence to evolve rapidly over the next decades, whereas the evolution of intelligence in animals proceeds at a pace of thousands or hundreds of thousands of years. Human intelligence develops at an intermediate pace, mostly driven by culture, as humans have become much smarter over the past few thousand years through the development of practices such as literacy, mathematics, and organized education. Another possible development is amalgamation of human and machine intelligence in cyborgs.

Considering multiple benchmarks debunks a linear scale of intelligence that would assign a particular number to the intelligence of a species, for example, that chimpanzees have an IQ of 40. Moreover, nonhuman animals exceed humans in some intelligent abilities. In perceiving, humans are

Report Card 4.1

Evaluation of animals with respect to features

Feature	Bee	Octopus	Raven	Dog	Dolphin	Chimp
Perceiving	A	B	A	A	A	A
Problem solving	C	C	B	B	B	B
Planning	C	C	B	C	B	B
Deciding	C	C	C	C	B	B
Understanding	F	F	C	C	C	B
Learning	C	B	B	B	B	B
Abstracting	C	?	C	C	C	C
Creating	F	C	C	C	C	B
Reasoning	F	F	C	C	C	B
Feeling	?	C	B	B	B	B
Communicating	C	F	C	B	B	B
Acting	B	B	B	B	B	B

See text for the grading scheme. The question mark indicates uncertainty about the appropriate grade.

Report Card 4.2

Evaluation of animals with respect to mechanisms

Mechanism	Bee	Octopus	Raven	Dog	Dolphin	Chimp
Images	?	?	B	C	?	B
Concepts	C	C	B	B	B	B
Rules	C	C	B	B	B	B
Analogies	F	F	C	F	F	C
Emotions	?	F	B	B	B	B
Language	F	F	C	C	C	C
Intentional action	C	C	B	C	C	B
Consciousness	?	C	C	C	C	B

surpassed by electromagnetic sensing in bees, ultraviolet vision in ravens, smell in dogs, and echolocation in dolphins. Moreover, some problem solving is enhanced by physical capabilities not found in humans, such as the strength of chimpanzees, the flying of bees and ravens, and the swimming and prehensile penises of dolphins. Nevertheless, human minds possesses enormous advantages over animal and machine intelligence.

5 Human Advantages

Many of the supposed differences between humans and other animals have been discredited: souls, language, emotions, creativity, tools, and culture. The claim that only humans have souls has been challenged by the lack of evidence for souls even in humans and less plausibly by the theological expansion for pets to have souls. Full human language does not operate in other animals, but species such as apes, prairie dogs, and even bees use sophisticated forms of communication. Many animals have simple emotions such as happiness and anger, and some have richer emotions such as grief and jealousy. Animals display creativity in coming up with insightful solutions to problems and in developing new tools, like those used by chimpanzees, crows, and other species. Bands of chimpanzees display culture through different feeding practices acquired by social learning.

Nevertheless, psychologists and anthropologists have identified more specific characteristics that are unique to humans and provide advantages with respect to intelligent behaviors.

1. We seek harder and harder problems to solve and invent new methods to extend knowledge. We not only create new things, concepts, and hypotheses but also create new methods for creating new things, concepts, and hypotheses.
2. We explain the world using hypothetical causes and can even imagine impossible scenarios.
3. We can imagine and reflect on different situations concerning the far-off past and future.
4. We not only learn but can learn how to learn better. Cultural learning transmits skills across hundreds of generations.

5. We study our own species and others, developing general knowledge and comparisons.
6. We not only teach our own children but prolifically teach other people and even teach how to teach.
7. We build fires and cook our food.
8. We use tools to create other tools and extend the range of problems we can tackle.
9. We modify our environments and expand our habitats to extreme locations all over the world.
10. We navigate with physical maps.
11. We not only have social emotions that enable us to interact with each other (embarrassment, guilt, shame, gratitude, pride) but have emotions about emotions, e.g., fear of embarrassment.
12. We evaluate ourselves and other people. We reflect on own behavior and engage in moral reasoning and judgment about ourselves and others. We punish strangers for breaking rules.
13. We tamper with ourselves, fixing imperfections by means of glasses, hip implants, and surgery to change the odds of natural selection.
14. We communicate linguistically with richly interconnected syntax, semantics, and pragmatics.
15. We not only tell jokes but can make jokes about joking.
16. We collaborate with each other and risk our lives for ideals.
17. We consider the intentions and beliefs that produce actions. We can reason about our past, present, and future motives, beliefs, and actions.
18. We have modified societies by creating governments and laws.

None of these capabilities is found in animals, so we are pressed to know their mental and neural origins.

Following Suzana Herculano-Houzel, I propose that these advantages result from humans having primate brains with an unusually large number of neurons in cortical areas. Cooking enabled human ancestors to evolve big brains that support improved mechanisms for representing the world and making inferences about it. With enough neurons, brains can operate recursively, looping to have representations about representations about representations and even inferences about inferences about inferences. Such rich representations and inferences enable people to have languages and emotions that support social cooperation that leads to degrees of culture and technology not found in other species. The increasing pace of culture and technology continues to move the human species further beyond the capacities of other animals.

One of the great accomplishments of human culture and technology is the development of machines that approximate human intelligence. Reflecting on differences between humans and other animals also sheds light on the current status and future prospects of machine intelligence. In chapter 3, we saw that advances in artificial intelligence and robotics are bringing machines closer to humans. But out of the eighteen human advantages over animals I have listed, only number 8 (tools for making tools) and number 10 (navigation with maps) are accomplished by machines today. However, some animals also have capabilities that go beyond the reach of current AI, such as understanding causality and being conscious.

Because biological evolution moves so slowly, the gaps between humans and other animals are unlikely to close. But cultural and technological evolution moves rapidly, and no good arguments show that machines could *never* have any of the capacities in my list of eighteen differences. Explaining the human capacities that distinguish us from other animals will provide lessons about what AI needs to draw closer to what people can do. Whether we actually want to close the gap between natural and artificial intelligence is an ethical issue I examine in chapters 7 and 8.

Cooking Up Big Brains

Suzana Herculano-Houzel is a Brazilian neuroscientist who wondered in 2003 how many neurons humans have in their brains. The standard number in the textbooks was that people's brains have 100 billion neurons, but she was unable to find any experimental basis for this estimate. So she creatively came up with her own method, which she calls "brain soup." She figured out how to take rat brains, fix them in paraformaldehyde to make the cells stable, dissect the brains into areas such as the cerebral cortex, slice the brain parts into portions that dissolve into soup, stain the cell nuclei with fluorescent dye, then count the number of cells under a microscope. This process produced not only an accurate determination of the number of cells in the brains of rats but also the first reliable finding of the number of neurons in humans, which turns out to be around eighty-six billion. She has used the same method to count the number of neurons in the brains of numerous animals, including elephants and primates. She scandalized cat lovers by finding that cat brains have only about half as many neurons as dog brains.

In 2016, Herculano-Houzel published a fascinating book called *The Human Advantage*. She says that our first neural advantage is that we are primates, an order of species that evolved the capacity to pack a large number of neurons into a small brain. Compared to rodents, dogs, and elephants, primates get much more brain for comparable space. For example, chimpanzee brains have about the same volume as cow brains but have many more neurons that enable a more flexible behavioral repertoire.

Second, around 1.5 million years ago, the hominid ancestors of current humans made a cultural breakthrough that allowed them to develop much larger brains than had previously been possible. The problem with large brains is that they require a lot of energy to keep billions of neurons firing. At around 1.3 kilograms, or 3 pounds, human brains are only about 2 percent of the size of a body that weighs 68 kilograms, or 150 pounds. But human brains consume around 20 percent of the total energy supply gained from foraging for food. Adding more brain requires finding more food, which is limited by the number of daily hours when an organism can be looking for things to eat.

The cultural breakthrough that solved this problem for our ancestors was cooking, which makes both animal and plant products much more digestible. Early humans had already acquired the ability to walk upright and to hunt and gather collectively. But sometime around the transition from *Homo habilis* to *Homo erectus*, people also figured out how to use fire to cook accumulated food. This discovery-invention, along with the cultural transmission of methods for cooking, made acquiring enough calories to support lots of neurons much more feasible. Over the next million years, brain size increased rapidly, leading to the large brains of *Homo sapiens* at around eight-six billion neurons. Other apes such as orangutans and gorillas grew larger bodies but not larger brains, which they did not have sufficient energy sources to support.

The third major neural advantage of human brains is that the cerebral cortex is proportionately much larger than in brains that are overall still bigger, such as those of elephants and large whales. The cortex is responsible for the sophisticated kinds of representation and inference needed for the capabilities listed in my eighteen advantages. Animals generally get smarter as their brains get larger, but in chapter 4 we observed that the correlation is not perfect: elephants and some whales have several times more neurons than humans but exhibit much less intelligence. The major difference is that the vast majority of their neurons are concerned with controlling their large bodies, with a much smaller percentage of their brains dedicated to cortex. The big question is: what does the cortex do that makes people so much smarter than elephants, whales, dolphins, and other animals? I aim to connect Herculano-Houzel's three neural advantages to my eighteen performance advantages.

The Looping Mind

How do large brains enable people to carry out their most advanced mental functions? Michael Corballis is a New Zealand neuroscientist who proposed a compelling answer: what distinguishes us in the animal kingdom is our capacity for recursion, the ability to embed our thoughts within other thoughts. In more familiar language, our brains are capable of looping effects in which we have thoughts about thoughts and other mental representations about representations. For example, you can plan to make better plans and worry about feeling depressed.

Corballis drew his own example of recursion in figure 5.1 as a thinker thinking about a thinker thinking about a thinker. You can create your own example of visual recursion by standing in front of a large mirror while holding a small mirror that generates pictures of you repeatedly. Verbal examples of recursion include the children's song about an old lady who swallowed a goat to catch a dog that she swallowed to catch a cat that she swallowed to catch a bird that she swallowed to catch a spider that she swallowed to catch a fly that she swallowed for no known reason. Among my favorite examples of recursion are five-word sentences in which each word is the same: "Buffalo buffalo buffalo Buffalo buffalo" and "Police police police police police." The first of these can be interpreted as saying that animals from the city of Buffalo intimidate animals from the city of Buffalo. A classic philosophical example is the paradoxical sentence "This sentence is false," which is true if it is false and false if it is true. Recursion can also be important for physical actions in the world, as in the human ability to make tools that make tools. For example, blacksmiths make tools such as axes using tools such as anvils and hammers that were made by other tools.

Figure 5.1
Recursive thinking about thinking. From M. C. Corballis, *The Recursive Mind: The Origins of Human Language, Thought, and Civilization* (Princeton, NJ: Princeton University Press, 2011), 2. By permission of Princeton University Press.

Corballis argues that recursion is responsible for the great power of the human mind. He rejects Noam Chomsky's claims that recursion is an essential feature of all human languages, because a few languages do not use recursion. But Corballis recognizes that almost all human languages employ various degrees of embedding and self-reference. In contrast, the kinds of animal communication described in chapter 4 are limited to strings of motions, gestures, or symbols (in the case of trained chimpanzees and bonobos) that do not exhibit recursion. Humans are able not only to tell jokes but to make jokes about jokes (meta-jokes), as in "A priest, a rabbi, and a leprechaun walk into a bar, and the leprechaun says, 'Begorrah, I think I'm in the wrong joke.'"

Corballis also describes the importance of recursion for the ability of people to think of themselves both in the past and in the future. You can connect your current self with your past self, for example, when you remember the last time you went to a city that you are now visiting. Even more powerfully, you can imagine yourself in the future visiting a city that you have never gone to before. In both cases, you are thinking about yourself doing things, and you can even think about what you are thinking in the past or what you expect to be thinking in the future. Animals have some cases of long-term memory and imagination, for example, when jays remember where they stored different kinds of foods, and when chimpanzees stockpile rocks to throw at spectators. But only humans can accomplish layers of looping thoughts about the past and future.

Moreover, you can think not only about your own thoughts but also about the thoughts of other people. For example, you can imagine how your friend would react to a restaurant that you want to suggest. Many tiers of looping are possible, as when you imagine what your friend will think about what you will think if the friend does not like your suggestion. Humans can use general knowledge and personal experience to make sometimes accurate inferences about what other people are thinking, including what other people are thinking about them.

Empathy is an important form of human interaction that requires both cross-temporal imagination and thinking about what other people are thinking. For example, if you have a friend who has lost a family member, you can imagine how your friend is feeling by remembering how you felt in the past in a similar situation, and project onto your friend how he or she is feeling now and will probably feel in the future.

Humans are capable of complex emotions, including recursive ones. Nested emotions are emotions about emotions. You can be happy that you are happy, or sad that you are sad. Some of your desires are ones that you desire to have, for example, wanting to work harder or to be nicer to your family. But other desires are ones that you desire not to have, for example, a craving for chocolate cake or lust for a neighbor's spouse. People can also have fear of embarrassment, hope for forgiveness, love of honor, longing for love, fear of fear itself, lust for glory, fear of shame, hatred of boredom, wanting to be brave, falling in love with love, fear of commitment, disgust at lust, daring to be proud, pride of love, wishing for trust, anxiety about attachment, and fear of disappointment.

In a blog post where I introduced the concept of nested emotions, I speculated that only one level of nesting might be possible in human brains, but commentators offered examples with more nesting:

I feel aggravated about my fear of attachment.

I feel embarrassed about my hope for love.

I feel guilty that I enjoyed hating attachment.

Such nesting of emotions requires repeated recursion.

All the eighteen human advantages depend on looping effects. Their importance is obvious in advantages such as learning to learn and making tools that make tools. Even the less obvious cases of cooking and navigation require looping so that people can imagine themselves in the future as preserving fire for cooking and as getting to places with better supplies of food and shelter. During the past ten thousand years, humans have learned to learn by using many inventions, including writing, reading, mathematics, schools, printing, computing, and the internet.

The Looping Computer

Recursion is a natural property of programming languages, for example, in this definition of a procedure for calculating the number of elements in a list:

Define length-of list

If list is empty, then return 0.

Otherwise add 1 to length-of the list with the first element removed.

This is recursive in that the definition of the function length-of refers back to itself, but it yields a definite answer because it keeps shortening the list.

Although computer programming languages are capable of recursion and other forms of iteration, most computer programs lack the forms of recursion and embedding that are needed to support intelligence. For example, my favorite programming language, Lisp, elegantly employs recursion, but typical Lisp programs do not use the looping effects needed for language, imagination, and understanding of other minds. We need to look at specific computational models to evaluate the extent to which they support higher levels of recursion.

IBM Watson can support recursion because it has powerful methods of natural language processing, but it is much better at syntax than semantics. The power of human language comes from the wonderful mesh between recursive syntax and recursive semantics, where meanings build on meanings in step with the embedding of syntactic structures within syntactic structures. For example, consider the sentence "The boy chased the dog who was chasing the boy." The sentence embeds the clause "who was chasing the boy" as a syntactic operation, but it is also crucial that the sentence carries the meaning that the boy who was doing the chasing is the same as the boy being chased. Watson is excellent at syntax but limited in meaning that connects to the world.

Because self-driving cars have interactions with the world that include learning sensory patterns, they are much better at world-connected semantics than Watson. Such cars can also self-reference by encoding information about car identity, for example, about Waymo 196. But the Bayesian networks and deep learning neural networks used in self-driving cars lack the syntactic capabilities to produce complex representations with embedded structures, for example, expressing the relation that Waymo 196 thinks that Waymo 233 thinks that Waymo 196 is driving erratically. Similarly, the deep learning neural networks that have made AlphaGo and similar programs successful do not have enough internal structure to allow for self-reference and embedding of representations about representations. Other neural networks discussed later do have these capabilities.

Google Translate has no difficulty coming up with a good French translation of "the boy chased the dog who was chasing the boy": "le garçon a poursuivi le chien qui poursuivait le garcon." But we saw that Google Translate works statistically by finding corresponding sentences using large databases, with little understanding of syntax and absolutely none of semantics. So the recursiveness of Google Translate is an artifact of the recursiveness of human language. Similarly, the ability of some 3D printers to make parts for themselves depends entirely on human programmers without involving any degree of self-representation.

Alexa and recommender systems similarly lack the looping capabilities of human syntax and semantics. Humans are also recursive with respect to the pragmatics of language, for example, in our ability to consider and accomplish goals that accomplish goals right up to our most fundamental needs. For example, my goal today is to write a thousand words that contribute toward my goal of writing this draft, which contributes toward my goal of writing this book, which contributes toward my need for achievement. The closest performance to such goal chaining is a New Caledonian crow getting food by following a series of eight steps that include stones, sticks, and string. But no evidence suggests that crows can think of themselves as performing these steps; unlike magpies, crows fail the mirror test for self-recognition.

Some artificial intelligence systems are capable of chaining goals to accomplish more goals, going back to Newell and Simon's Human Problem Solver of the 1970s. It would be a mistake to conclude that artificial intelligence is impossible based on current limitations with respect to recursive syntax, semantics, and pragmatics. Unlike beasts, bots are continuously improving, and researchers may be able to surmount the considerable engineering challenge of making computers capable of all the kinds of recursive thinking that contributes so much to human intelligence. Already we have computer programs such as Yacc (yet another compiler-compiler) that help to make new programming languages that can be used to write novel programs. In contrast, even with new techniques of genetic engineering such as CRISPR, there are no foreseeable prospects of enhancing crows, chimpanzees, or other animals to match humans in looping abilities.

How Brains Became Recursive

Now we have a big gap to fill in explaining the origins and operation of human intelligence. Natural selection of primate brains and the development of cooking explain how human brains got so big, and the capacity for looping representations explains how humans became capable of imagination, understanding other minds, and language. But how did big brains acquire recursiveness? One possibility, proposed by Robert Berwick and Noam Chomsky, is that a simple genetic mutation enabled brains to merge two objects into a new object that contains both of them. This proposal is short on details about the relevant genes and about the neural processes that were modified.

A richer explanation of the origins of mental loops comes from the exciting new field of theoretical neuroscience. Modern neuroscience began in the late nineteenth century, when the Spanish biologist Santiago Ramón y Cajal realized that newly discovered cells in the brain, later called neurons, were the major parts in the brain's mechanism for thought. The twentieth century brought major advances in experimental techniques for investigating brains, including recording signals from single neurons and less invasive brain scans. Brain-scanning techniques such as functional magnetic resonance imaging gave rise to the subfield of cognitive neuroscience, which looks at the neural correlates of human behavior such as problem solving.

Experimental data provide useful descriptions but not explanatory understanding, which requires identifying the causal mechanisms that produce what is observed. Since the 1980s, the field of theoretical neuroscience has risen to build mathematical and computational models that explain experimental results. Theoretical neuroscience is only a few decades old but has made major advances in figuring out how groups of neurons interact to produce intelligent thought.

My colleague Chris Eliasmith has developed new theories of neural representation and processing that bridge the gap between large brain size and recursive operations. Start with a single neuron that can begin to represent things in the world by firing in different patterns. For example, a neuron connected to the visual system might send electrical signals to other neurons when its inputs indicate that something blue has been observed. More powerfully, over a period of time such as a second, the neuron might fire rapidly rather than slowly, or it might fire with a particular temporal pattern such as fire-fire-rest-fire.

One neuron by itself cannot represent much, but a whole group of neurons working together can have firing patterns that are capable of representing shades of blue or lines in different orientations and shapes. A harder task would involve representing something that is blue and made up of four lines at right angles forming a blue square. For the brain to represent a blue square, its neurons must combine representations of blue, lines, and angles into a single configuration.

The operation by which neurons combine simpler representations into more complex ones is called binding, for example, the binding of blue, lines, and angles into a blue square. All animals, even ones with small numbers of neurons such as bees, are capable of this basic binding, for example, when a bee has neurons that fire to represent the flowers with pollen in a particular location. This basic binding does not require any recursion and might be done by the neurons for flowers and the neurons for location firing in synchrony with one another.

Eliasmith showed mathematically how groups of neurons can produce not only bindings but also bindings of bindings, and even bindings of bindings of bindings. He calls the resulting neural representations *semantic pointers*, which bind together sensory and motor patterns into new patterns that can function as symbols. For example, when visual inputs about blue, lines, and angles get combined into a neural representation for a blue square, the result is a new neural representation that can then be bound into more complicated representations. The thought that the blue square is underneath a red circle can be built up by neural groups that represent the blue square, the red circle, and the relation *under*. This process can iterate with the production of even more complicated relations, such as that an artist moved the red circle to be on top of the blue square.

The mathematics of semantic pointers are well developed, but for the audience of this book, metaphorical descriptions are more appropriate. You can think of the brain's production of semantic pointers as being like baking a cake, where ingredients like flour, butter, sugar, and eggs are bound together into something new that can then be combined with other things like icing. The cake retains some of the characteristics of its ingredients, such as the sweetness of sugar. Alternatively, you can compare semantic pointers to a braided rope formed by twisting together smaller strands, resulting in a new, stronger object that can still be decomposed into its components.

The operations that construct semantic pointers can be linguistic, as in sentence formation, but you can also make a mental image of a red circle on top of a blue square, without using words. Eliasmith's semantic pointers have a property that I call "modal retention," which is not found in any other artificial neural networks. This term means that the pattern of firing that makes up a semantic pointer retains an approximation to the sensory and motor inputs that produced it. For example, the semantic pointer for a blue square continues in compressed form the pattern of firing for blue that went into it. Moreover, the motor operation of moving a shape can also be partially retained in the semantic pointer for moving a red circle on top of a blue square.

Semantic pointers have three aspects that are crucial for understanding the recursiveness of thought. First, they are multimodal, in that they can capture representations based on a wide range of sensory and motor modalities including vision, smell, touch, and bodily motion. This means that they can explain the nonverbal mental operations of animals, as well as the verbal combinations found in human language. Second, they perform modal retention, so that they do not throw away the sensory-motor information that goes into making them. This retention shows how world-connected semantics can approximately survive through all the bindings that go into producing more complicated semantic pointers.

Third, semantic pointers can be combined into more complicated semantic pointers, producing many tiers of recursion, such as a blue square that contains a blue square under a red circle. In Eliasmith's computational models, the bindings of sensory-motor inputs into semantic pointers and the bindings of semantic pointers into more complicated ones are all performed by neurons. This process does not allow an infinite amount of recursion, because people do not have an infinite number of neurons. People get lost after a small number of repetitions along the lines of "I thought that you thought that I thought that you thought that . . ." Nevertheless, people have far more repetition and recursion than can be accomplished by animals with smaller brains. Computer simulations show that it takes a lot of neurons to carry out the binding process, thus explaining why it was important that humans' discovery of cooking enabled them to develop larger and larger brains. Large brains with more neurons enabled more formation of semantic pointers and binding into more complicated semantic pointers.

Emotions require bindings because they combine representations of situations, physiological reactions to the situation, and appraisals of the situation. All emotions are about something, for example, a person that you love or a situation that makes you happy or sad. In addition, all emotions involve physical changes, for example, your heart starting to beat faster when you see someone of romantic interest or your stomach churning when you suddenly have to give a public speech. Emotions are not just physiology, however, because people have more nuanced emotions than bodily states capture. For example, fear and anger are physiologically similar, involving virtually identical patterns of heart rate, breathing, and other bodily changes. The difference between fear and anger is that they involve different evaluations concerning how a situation is affecting your goals: fear marks a threat to your survival goals, and anger is directed at a thing or person that is blocking your goals.

Emotions operating in brains require firings in groups of neurons that bind the representation of the situation, perception of physiological changes, and appraisal. All of these can operate in animals, where the representation, perception, and appraisal are performed without words. For humans and a few other animals who are capable of representing themselves, a fourth aspect can also operate: the sense of a self that is having the emotions, as when you know that it is you who is happy about meeting a friend. Emotions that are recursively nested require further layers of binding. In chapter 6, I examine further the question of animal emotions and consider the possibility of emotions in computers.

Several stages in the development of brains were required for the development of the kinds of recursion found in humans. At stage I, early brains became capable of binding sensory inputs into combined representations, such as blue line. At stage II, binding of bindings could produce representations of objects, such as blue square. At stage III, binding of bindings of bindings could produce more complicated representations, such as the blue square on top of the red circle. Finally, stage IV occurred around the evolution of *Homo erectus* into *Homo sapiens*, when binding became sufficiently powerful to support language, imaginings about the future, and explanations by hidden causes such as spirits and the mental states of other people.

To appreciate how simply adding more neurons could lead to the qualitative differences in recursion, we need an important concept called a critical transition. For example, when you fill an ice cube tray with water and put it in your freezer, the water gradually gets colder and colder while remaining liquid. But with enough time, it solidifies into ice, a qualitatively significant change resulting from quantitatively gradual changes. Similarly, adding more neurons in the brains of hominids led to the qualitative change that produced the looping mind liberated from the limitations of representations bound to sensory-motor systems. A difference in neural degree turned into a difference in psychological kind, like the character in Hemingway's *The Sun Also Rises* who went bankrupt gradually and then suddenly.

The six smart machines in chapter 3 and other AI systems lack the key features that connect big brains to recursiveness. The neural networks used in deep learning differ from the neural networks used in semantic pointers in several key respects. First, artificial neurons in deep learning represent by rate of firing, not by pattern of firing, and there are far more patterns than rates. If a neuron can fire 100 times per second, then there are just 100 rates at which it could be firing over the duration of the second, with 5 meaning slow and 100 meaning really fast. But the number of different firing patterns that the neuron could produce is 2^{100}, a lot more than 100. This greater representational capacity is important for capturing different features in the world and in the body's internal states. That huge number is only for one neuron and increases enormously when you consider the number of firing patterns in groups of thousands or millions of neurons.

Having spiking neurons is important for the second crucial feature of what makes semantic pointers different from other computational approaches: modal retention. To avoid throwing away information about the sensory-motor origins of higher-level representations, neurons need to represent many different states of affairs, and large numbers of neurons and large numbers of neural patterns make this happen. Even more important than the number of neural patterns is the property of modal retention that the neural representations contain an approximation to the sensory-motor signals that produced them.

The third crucial feature of semantic pointers not found in deep learning networks is their symbol-like capacity to get combined into new representations. Deep learning can produce rich representations of patterns in the world, and it is even capable of degrees of abstraction described in chapter 3. But it does not generate symbols that can be combined into other symbols in ways important for many kinds of creativity. For example, scientists combine existing concepts into new theoretical concepts such as *atom*, which was understood initially as an invisible, indivisible particle. Unlike neural networks trained by examples, semantic pointers can support the human advantages of going beyond what is currently observed to think about the future and hidden causes such as the mental states of other people.

None of the six smart machines in chapter 3 have the crucial properties of representation by spiking neurons, modal retention, and combination into progressively complicated representations. For artificial intelligence to make the critical transition to human-level intelligence, it will need to add such features that bridge the gap to recursiveness and meaning. However, this gap does not show that artificial intelligence is impossible, because Eliasmith already has powerful computer models in operation with tools available from his company, Applied Brain Research. These models operate on neuromorphic chips that implement millions of neurons, resulting in highly parallel, energy-efficient, and computationally effective operations.

So semantic pointers fill the gap in the story of the evolution of human intelligence between large brains and mental looping effects that produce major human advantages over bots and beasts. Figure 5.2 encapsulates the story, depicting how already sophisticated primate brains got bigger thanks to cooking. More neurons led to more recursive binding into neural representations that retained aspects of their sensory-motor origins while offering new capacities for combining into symbols. The result was looping capabilities for both verbal and nonverbal representations that generated human advantages such as complex language, tools to make tools, and social understanding. These advantages in rudimentary form increased the abilities of

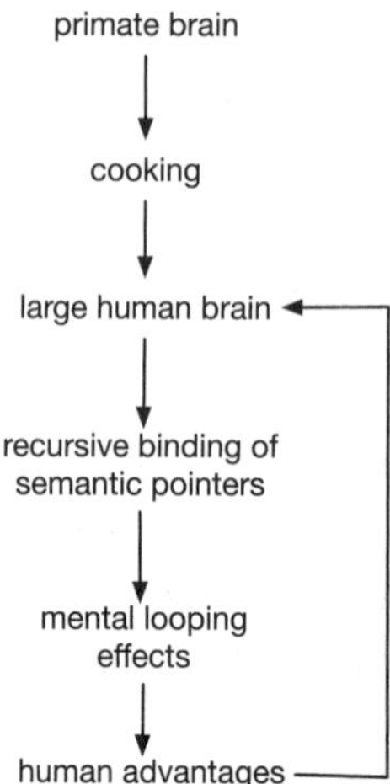

Figure 5.2
Origins of human advantages.

early humans to survive and reproduce, favoring the growth of even bigger brains, as shown by the arrow back from advantages to brain size.

Causality

Looping effects are not the only reason why human intelligence is still far superior to AI. Yoshua Bengio, one of the pioneers of deep learning, remarks:

> Researchers are trying to understand what the issues are, such as, why is it that we can't build machines that really understand the world as well as we do? Is it just that we don't have enough training data, or is it that we don't have enough computing power? Many of us think that we are also missing the basic ingredients needed, such as the ability to understand causal relationships in data—an ability that actually enables us to generalize and to come up with the right answers in settings that are very different from those we've been trained in.

Here animals have a huge advantage over machines. Cats know that if they claw at a mouse, then the mouse will move, and animals from octopuses to crows to chimpanzees appreciate that tools can be useful in manipulating the world. Humans needed an understanding of causality to use fire as a tool for making food more palatable: igniting a pile of wood causes a fire that causes food to cook and taste better.

The most sophisticated AI research on causality is by Judea Pearl, who has done brilliant work on causal diagrams, Bayesian reasoning, mathematical models, and counterfactuals. But in his most recent book, he characterizes causation as follows:

> A variable *X* is a cause of *Y* if *Y* "listens" to *X* and determines its value in response to what it hears.

This definition is uninformative for three reasons. First, the terms "listens" and "hears" are vague, metaphorical, and misleading: no listening is going on when cancer cells are caused to mutate by cigarette smoke. Second, the definition is circular, for the word "determines" is just another version of "causes." Third, the definition assumes that causation is a relation between variables, but in ordinary life, causation is usually taken as a relation between events, for example, when the event of pressing on a car's gas pedal causes the event of the car speeding up.

We should not be surprised by the failure to get a good definition of causation, because definitions outside of mathematics always fail for reasons that include vagueness, circularity, and counterexamples. A better way to characterize concepts is to use the method applied to intelligence in chapter 1, analyzing the concept *cause* by using standard examples, typical features, and explanations. Standard examples of causes include pushes, pulls, motions, collisions, actions, and diseases. Causes typically but not universally have features: temporally ordered events with causes before effects, sensory-motor-sensory patterns, regularities expressed by rules, manipulations and interventions, and statistical dependencies with causes increasing the probabilities of effects. Finally, causes serve to explain why events happen and why interventions work.

Everyone is familiar with standard examples of causation, such as pushing a door open. Causes typically precede their effects, even though we can imagine traveling backward in time. Another typical feature of causality is that it exhibits sensory-motor-sensory patterns, where a perception followed by an action leads to a different perception. Infants can recognize such patterns when they see rattles, hit them with their hands, and then see them move and hear them make sounds. Similarly, nonverbal animals observe sensory-motor-sensory patterns, as when a chimpanzee sees a rock and a nut, hits the nut with the rock, and then sees, hears, and feels the nut split open. Other sensory-motor-sensory-patterns operate when a chimp sticks a twig in a termite hole and pulls it out covered with termites, or when a crow drops a stone in a tube of water and sees the water rise.

Language and mathematics provide an enhanced understanding of causality through concepts like regularity and statistical dependence. But these depend on the sensory-motor-sensory patterns already familiar to preverbal infants and animals. The additional features of causality complement but do not replace the original sensory-motor understanding. Consciousness is an important contributor to this understanding because it enables the simultaneous consideration of sensory and motor components, as when an infant is aware of a rattle-movement-noise combination.

Pearl deals with the problem of the difference between causation and correlation by distinguishing between the probability of *X* given *Y* and the probability of *X* given *doing Y*. This distinction is useful because it captures the aspects of manipulation and intervention that are often considered an important aspect of causation, but Pearl provides no semantics for the *doing* operation. From the point of view of language and mathematics, this omission is serious, but from a broader cognitive perspective, we can attribute understanding of doing to the sensory-motor experiences that people share with animals.

Computer programs do not naturally have these experiences, although potentially they could acquire them through sensing devices and processing. Such resources go beyond what deep learning can do by training from examples that involve no motor operations on the world. Instead we need a robot that is capable of perceiving, acting, and noticing the results of the perception-action pairing, from which it could learn sensory-motor-sensory patterns.

Potentially, semantic pointers could be used to create multimodal rules of the form *if sensation and motion then sensation*. For example, a learned rule might be *if <rattle is still> and <hand hits rattle> then <rattle moves>, and <rattle make sounds>*. In this rule, I have placed the crucial parts of the if and then clauses in brackets to indicate that they are not verbal representations. Rather, they are the visual, motor, and auditory representations that operate in the brains of chimpanzees, infants, and neural networks that accomplish modal retention.

Without such sensory-motor-sensory patterns, all current AI programs are deficient in their understanding of causality. Symbol-based programs can dodge this limitation by using the word "cause," but that is just syntax without semantics. Watson, self-driving cars, and other deep learning systems are similarly deficient in causal understanding compared to humans and other animals. Until computational devices become capable of sensory-motor appreciation of causality, they lack the kinds of causal understanding that are important for explanation and intentional action.

Causes are also important for learning because they help to focus on what matters to accomplish the goals of the learner rather than on the huge number of possibly relevant variables that go into training a deep learning network. Layers of hidden causes introduce looping effects, for example, when biochemistry causally explains DNA operations that explain genetic inheritance that helps to explain biological evolution.

I am not saying that such understanding is impossible for machines. Computational models employing semantic pointers are already running on neuromorphic (brain-like) chips and controlling simple robots. In principle, they are capable of being programmed with the sensory-motor-sensory schemas that I have claimed are basic to fully appreciating causality. It would be more elegant if they could learn such schemas, but that is too much to expect when it is plausible that schematic causality is innate in animals: infants exhibit it within a few months of being born, and some animals exhibit the ability to get around in the world immediately after birth.

Causality and recursion work together to enhance the mental powers of humans. Causality is recursive in two respects. First, it displays iteration as in the following poem:

> For want of a nail the shoe was lost.
> For want of a shoe the horse was lost.
> For want of a horse the rider was lost.
> For want of a rider the message was lost.
> For want of a message the battle was lost.
> For want of a battle the kingdom was lost.
> And all for the want of a horseshoe nail.

Second, causality can feed back on itself, for example, when a husband insults a wife, who then insults the husband, who then insults the wife again. Causal feedback loops are a dangerous part of climate change, when global warming melts permafrost that emits methane that increases global warming. Animals whose understanding of causality is linked to perceptual objects such as rocks and nuts cannot handle recursive causality because they lack enough neurons to produce bindings of bindings needed to represent causes of causes.

Animals also miss out on causality that comes from hidden, nonobservable causes such as atoms and gravity. Sensory-motor-sensory schemas only apply to observed events, but humans are capable of looking for underlying causes that go beyond what they can perceive. For example, Homer's *Iliad* is one of the oldest literary texts and begins with the following paragraph, in which I italicize hidden causes:

> Sing, O *goddess*, the anger of Achilles son of Peleus, that *brought* countless ills upon the Achaeans. Many a *brave soul* did it send hurrying down to *Hades*, and many a hero did it yield a prey to dogs and vultures, for so were the counsels of *Jove* fulfilled from the day on which the son of Atreus, king of men, and great Achilles, first *fell out* with one another.

This text assumes the existence of gods and goddesses, along with an underworld where people go when they die. It also mentions the psychological dispositions to anger, bravery, and falling out that underlie but go beyond explicit behavior. More evidence that the ancient Greeks were adept at considering hidden causes comes from the doctrine of atoms proposed by Democritus and the concepts of *aether* (the material that holds up the stars) and *psyche* (soul) proposed by Aristotle.

Deep learning has no way of generating such hidden causes, but some computational systems already incorporate relevant mechanisms. These include abductive inference that generates hypotheses, conceptual combination that generates new concepts from old ones, and analogy that generates hidden causes by considering similar situations. Analogy has contributed to scientific discovery, for example, when the ideas about water waves inspired ideas about sound waves that later inspired ideas about light waves. Animals besides humans cannot aspire to hidden causes, but artificial intelligence can if it goes beyond the mechanisms implemented in the six smart machines in chapter 3.

Social Advantages

Intelligence is usually taken as a property of individuals, but most human accomplishments take place in groups such as families, schools, laboratories, companies, and governments. Psychologists have shown how emotional and social intelligence enables people to understand and interact with others to solve problems that are more complex than one person could tackle alone. For example, the contents of science journals such as *Nature* show that almost all scientific research today is collaborative, enabling people to combine their knowledge and skills.

Some animals demonstrate social coordination, such as wolves hunting together and male dolphins collaborating to control females. But the degree and nature of collaboration by humans far exceed what is found in other animals. These differences result from recursive brains that are capable of language, understanding other minds, and foreseeing the future.

How Human Social Coordination Is Better

In his 2019 book *Becoming Human*, Michael Tomasello summarizes decades of research on differences between social coordination in human children compared to chimpanzees. Most of humans' remarkable achievements are based on how people cooperate. Some great apes are capable of understanding each other's minds, communicating intentionally, learning socially, hunting in groups, forming alliances, helping one another, and reciprocating actions. But Tomasello maintains that humans have unique forms of cultural coordination and transmission of information that provide our species with a huge social advantage.

Tomasello identifies three sets of processes that differentiate human development from that of other animals. The first set concerns the maturation of children, who become capable of forming intentions jointly with another person around nine months of age and become capable of forming intentions collectively with groups of people around three years. The second set concerns children's sociocultural experiences, which include being taught by adults and interacting in coordination with peers. The third set concerns self-regulation, in which children acquire the ability to monitor the perspectives and evaluations of social partners and use this information to control their own behavior.

Tomasello recognizes that our nearest ape relatives have skills for understanding, predicting, and manipulating their physical and social worlds. But he thinks they lack the ability to use joint attention, conventional communication, and pedagogy to work together as effectively as humans can. He maintains that what appears to be cooperative hunting by chimpanzees and other animals is just individualistic coordination in which each hunter is attempting to capture the prey for itself. In contrast, he thinks that early humans around 400,000 years ago began to obtain most of their food through active collaboration. Tomasello describes the biological advantages of becoming capable of socially recursive inferences based on the ability to recognize the intentions and attentions of others.

Moreover, early humans benefited by acquiring new motivations and emotions, including motives to work with others toward cooperative goals and the ability to feel sympathy for others along with expectations of fairness. Frans de Waal and others have described instances of concern with fairness and empathy in animals, but Tomasello insists that such concern is more intensely built into humans than into apes. Young human children are much more inclined to spontaneous collaboration. Chimpanzees are capable of some degree of self-regulation, for example, in deferring eating a small amount of food to gain more later. But their thinking is not regulated by anticipating how others would understand and evaluate their actions and thoughts. People excel at social recursion—thinking about how others think about what they are thinking. Societies thrive on such reflection through religions concepts such as *guilty conscience* and legal concepts such as *mens rea*.

Teaching is rare among animals, occurring only in a few species, such as meerkats and killer whales. But it is ubiquitous among humans, and all responsible parents teach children how to operate in their physical and social worlds. Through their parents and schools, children learn to adjust to their culture, following norms of conformity, sympathy, fairness, and other normative standards of their families and societies. Humans not only teach but teach how to teach in schools of education, the best of which rely on collaborative research concerning effective learning.

Teaching requires looping effects where the teacher knows what the student does not know and what the teacher wants the student to know. Kevin Laland conjectures that language evolved to teach close relatives about useful cultural variants including tools, foraging techniques, social signals, rituals, and medicines. Figure 5.3 supplements figure 5.2 to make more explicit the interaction of cognitive and social advantages such as cooperation and teaching.

Why Human Social Coordination Is Better

Explaining why people have these superior skills of social coordination requires specifying the nature and origin of the neural and psychological mechanisms behind the skills. The social recursion described by Tomasello is another case of the looping properties of minds. Larger brains with more

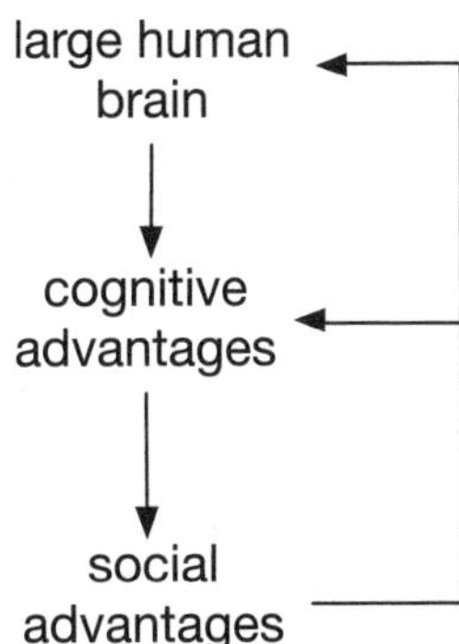

Figure 5.3
Interaction of brain size and cognitive and social advantages.

layers of binding led to the recursive thought that supports language, inferences about minds, and consideration of the future. Tomasello speculates that genetic changes led to the improved ability of humans to take into account joint intentions. Another possibility is that just having more neurons and binding capacity was enough to give early humans the ability to think more recursively about their own thoughts and the thoughts of others.

Brain size made humans more capable of thinking about other people, but alone it would not suffice to make them more capable of *caring* about other people. Animals display some forms of caring, such as raven couples raising their infants, and monkeys expressing sympathy and fairness. But human caring is much more expansive, as shown by institutions such as charities, hospitals, and welfare programs. Thanks to our large brains and social motivations that include caring, humans are capable of carrying out cooperative tasks that go far beyond the capability of other animals. Can machines cooperate?

Machine Social Coordination

Artificial intelligence machines are usually individual computers, but groups of computers are also capable of social coordination. The largest example of computer cooperation is the internet, which in 2019 was estimated to connect more than twenty billion computers and other devices. A long-standing research field in computer science goes under the heading of multiagent systems, also known as distributed artificial intelligence and decentralized intelligence. This field investigates systems composed of intelligent computational agents that compose artificial social systems, covering various kinds of communication and cooperation including group decision making.

Of the six smart machines discussed in chapter 3, only one operates as a multiagent system. Ongoing research investigates various ways in which self-driving cars can communicate and cooperate to keep traffic moving smoothly. When the vehicles communicate with each other, they can cooperate to avoid collisions and traffic slowdowns that occur when drivers adapt poorly to changes in road conditions.

Multiagent systems such as driverless cars show that social coordination already exists among machines, but how does it compare with human social capacities? You might think that computers would be adept at communication because they can share electronic information rapidly through wired or wireless connections. But such rapid communication assumes that the formats of communicating devices are the same.

IBM Watson should have no trouble communicating with another installation of IBM Watson, but the representational formats used by Watson and those used by self-driving cars have little in common. The vehicles produced by different companies have different data structures and algorithms operating in their programs, so there is no guarantee that a car built by Waymo could communicate in real time with a car built by Ford. The internet works well because of a common protocol governing the interactions of all participating computers and devices, and research is under way to establish a protocol (called V2X for "vehicle to everything") that would allow a vehicle to communicate with other vehicles and entities.

Tomasello describes how evolution gave people the means to cooperate and also the motivations to do so. Because computers today lack emotions, they do not have the goals that motivate people to work together. Computers can have purely cognitive goals in the sense of representations of future states of affairs that their algorithms are designed to accomplish, but they are currently incapable of caring about such accomplishments. Even more remarkably, they are incapable of caring about other computers with whom they need to work in accomplishing common goals.

Solitary animals such as octopuses and orangutans are individualistic, just out to accomplish their own goals. More social animals such as ravens, wolves, and chimpanzees are genetically inclined to operate in groups. Through biological evolution and cultural development, humans have acquired powers of coordination to accomplish shared goals. For computers to be able to accomplish better social coordination, they will need to get closer to human abilities to infer what is going on in other computers. For example, if one self-driving car detects another self-driving car that is performing erratically, it could come up with a mechanical explanation of what is wrong with it.

Such inferences are analogous to what people do when they identify what is wrong with the thoughts and emotions of people with whom they are trying to cooperate. They would be facilitated by the development of full language processing in self-driving cars and other multiagent systems, combined with body-based understanding of causality tied to sensory-motor-sensory schemas.

Wanting Humans to Be Superior

A 2018 article by an anthropologist and a primate researcher asked the question: why do we want to think humans are different? The authors provided two answers to the question, based on religion and on motivations to use animals for human purposes. Religions such as Christianity, Judaism, and Jainism have provided theological justifications for seeing humans as superior. But such justifications vanish with a scientific perspective that does not recognize a god who gave people dominion over animals. Independent of religion, people are motivated to see animals as different to justify using them for purposes including food, clothing, medical experiments, and entertainment. Such motivations should not be allowed to block critical assessment concerning factual questions about the abilities of animals or objective ethical assessment of how animals should be treated.

However, it is legitimate to ask the question in the other direction: why do we want to think that humans are *not* different? People have various emotions for wanting to minimize the differences between humans and other animals. Darwin insisted that the difference between humans and animals with respect to their mental faculties is only a matter of degree, rather than kind, because he wanted to forestall the objection to his theory of evolution by natural selection that humans are special because of our mental abilities. Pet owners and other people who are emotionally attached to animals are motivated to think of them as members of their families, worthy of love and affection like their children. Some animal researchers are motivated by the desire to avoid the diminishment or extinction of the animals whose study fascinates them.

Deliberations about the factual differences between humans and other animals should be based on evidence rather than motivational distortions from either direction. Accordingly, in chapter 1, I used the Attribution Procedure to give an objective basis for deciding what mental attributes are found in humans but not in other animals or in machines.

Although many supposed differences between humans and other animals have been discredited, I identified eighteen human advantages noticed by researchers with no taint of the religious or practical motivations to find differences. Humans are unusual in our abilities to learn to learn, teach to teach, make tools that make tools, make plans about the far future, develop theories concerning underlying mechanisms, use looping language, and so on. These advantages are not a killjoy list, for I acknowledge that animals have remarkable capabilities for language, emotions, creativity, tools, and culture.

In this chapter, I have noted these differences and explained how they came about. Primate evolution enabled humans to develop large brains, and cooking allowed brain enlargement to take off around 1.5 million years ago. Having more neurons organized into association areas in the neocortex enabled human ancestors to go beyond perceptual experience by producing bindings of bindings of bindings that generate representations of representations of representations. These recursive properties of brains enable looping properties of the mind, including the embeddedness of language, the use of complex analogies and causal reasoning, causal explanations of how minds work, imagination about the future, and productive collaborations with other people.

Consideration of these advantages also sheds light on why human minds are superior to current machine intelligence, which is also limited in its looping abilities. Animals would need millions of years of biological evolution to catch up to *Homo sapiens*, but artificial intelligence is developing rapidly, and there are no absolute impediments to computers advancing to overcome the human advantages. However, the technological challenges are enormous in both hardware and software to enable computers to match humans with respect to characteristics such as emotions and consciousness, and we should not expect the gaps to be filled anytime soon. More likely, science and engineering will eventually work around these gaps to produce machines that get the looping capacities of human beings, unless humans decide to handicap them to prevent machine ascendance, as I consider in chapter 7.

So far, my applications of the Attribution Procedure have been informal, but controversial questions about animal and computer minds require more systematic investigation. Next I delve into controversial issues that range from bacterial minds to computational consciousness.

6 When Did Minds Begin?

A provocative book by Arthur Reber argues that bacteria are conscious and the origins of mind are found in the simplest, single-celled organisms that arose billions of years ago. I think that minds began much later, and I contemplate alternative answers to the question of when consciousness originated.

1. Consciousness has always existed because God is conscious and eternal.
2. Consciousness began when the universe formed around 13.7 billion years ago, in accord with the doctrine of panpsychism, which says that everything has a bit of mind in it.
3. Consciousness began with single-celled life around 3.7 billion years ago (Reber).
4. Consciousness began with multicellular plants around 850 million years ago.
5. Consciousness began when animals such as jellyfish got thousands of neurons around 580 million years ago.
6. Consciousness began when insects and fish developed larger brains with about a million neurons (honeybees) or ten million neurons (zebra fish) around 560 million years ago.
7. Consciousness began when animals such as birds and mammals developed much larger brains with hundreds of millions of neurons, around 200 million years ago.
8. Consciousness began with *Homo sapiens* around 200,000 years ago.
9. Consciousness began when human culture became advanced around three thousand years ago (Julian Jaynes).

10. Consciousness does not exist, as it is just a scientific mistake (behaviorism) or illusion.

I find answers 6 and 7 more plausible than the alternatives, but the basis for selection should go beyond an intuitive hunch. The Attribution Procedure outlined in chapter 1 provides a systematic way of adjudicating the origins of mind. We need to examine claims that entities have consciousness and other aspects of mind, taking into account relevant evidence, alternative hypotheses, and underlying mechanisms.

I will use the Attribution Procedure to answer the following questions:

1. How do you know that you have a mind and that other people do too?
2. Are bacteria and plants conscious?
3. Do fish feel pain?
4. Do cats and dogs experience jealousy?
5. Can apes and other animals think analogically?
6. Will computers have minds?

Questions 2 through 5 are ongoing scientific controversies involving reputable researchers on both sides, and I hope that using the Attribution Procedure will help to establish reasonable consensus.

My Mind and Yours

I have no doubt that I have a mind and that the people reading this chapter also have minds. But I begin with these philosophical questions to illustrate the use of the Attribution Procedure on relatively simple cases. How do you know that you are conscious? One piece of evidence is that you feel you are conscious, but this might be a mistake, as behaviorist psychologists and some philosophers have argued. Fortunately, additional evidence indicates that you are conscious, including your verbal reports of conscious experiences and your complex behaviors such as ones related to pain, emotions, and imagery that can be explained by your having these conscious experiences. Moreover, we are beginning to discover deeper neurological explanations of how consciousness comes about through the interactions of numerous brain areas. So inference to the best explanation supports the hypothesis that you are conscious as superior to the alternative hypothesis that you only act as if you are conscious.

Figure 6.1 diagrams the good reasons I have for believing that I have a mind. The alternative hypothesis is that I am mistaken that I have a mind. Descartes pointed out in the seventeenth century that he could not doubt that he was thinking because doubt is a kind of thinking, leading to his famous conclusion about his own existence: I think, therefore I am. The Attribution Procedure is fuller because it displays more evidence that I have a mind, along with scientific knowledge about how minds result from brain activity. The hypothesis that I have a mind explains observations that include all of the twelve features of mind listed in chapter 1. That I have a mind is in turn explained by my having a brain in line with thousands of findings in cognitive neuroscience. The competing hypothesis is just that I am mistaken about having a mind, but this hypothesis totally lacks supporting evidence or explanation of why it might be true. So I can conclude that I have a mind beyond a reasonable doubt.

The same form of reasoning supports my conclusion that other human beings are also conscious. I do not have direct access to the experiences of others, but I can observe their reports and behaviors related to problem solving and the other eleven features of intelligence. Moreover, I know from neuroscience that the brains of other people are very similar to mine. Alternative explanations, such as that other people are mindless automatons, have no evidence to support them. Therefore it is plausible that other people have minds just as I do. This is not just a weak argument from analogy but an inference to the best explanation that relies on the fact that the evidence and explanations for the consciousness of other people are almost as convincing as the arguments for myself. Figure 6.2 diagrams the

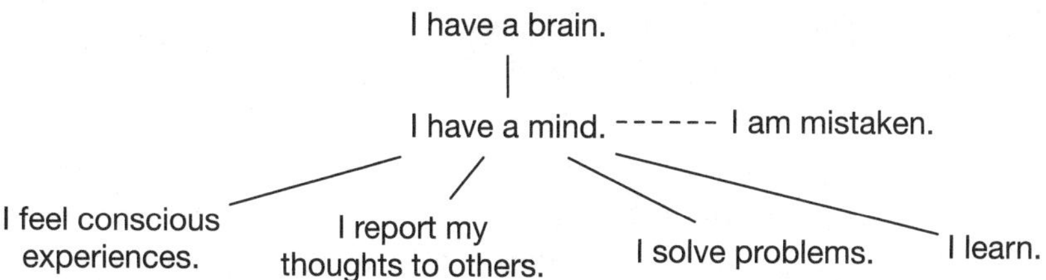

Figure 6.1
Diagram for attributing mind to myself. The solid lines indicate coherence resulting from explanations, and the dotted line indicates incoherence between competing hypotheses.

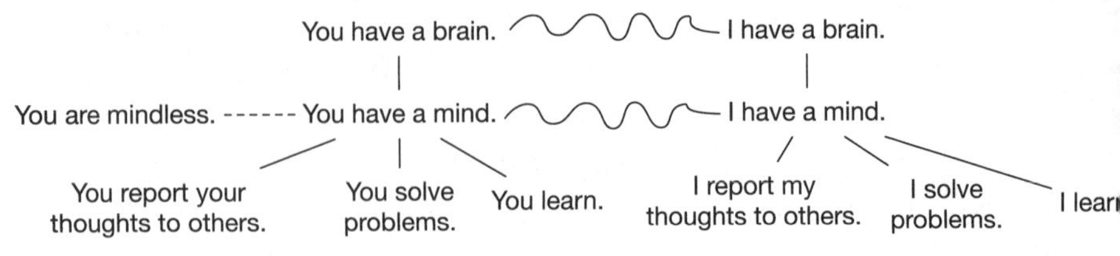

Figure 6.2
Diagram for attributing mind to others. The solid lines indicate coherence resulting from explanations, and the dotted line indicates incoherence between competing hypotheses. The wavy lines indicate coherence resulting from analogy. There should also be wavy lines between the analogous pieces of evidence, for example, between *You learn* and *I learn*.

attribution of minds to other people, including the analogy between how my mind explains my behavior and how your mind explains your behavior. The result is not as indubitable as my conclusion that I have a mind, but still holds beyond a reasonable doubt.

Do Bacteria and Plants Have Minds?

The evidence for consciousness in nonhuman animals such as cats and dogs is weaker because they cannot report their conscious experiences. The evidence becomes still sparser if you move down to smaller-brained animals such as bees and fish. Honeybees do exhibit reward-related behaviors, and fish exhibit pain-related behaviors, but these may not require an explanation based on conscious experience. Simpler animals such as jellyfish and even plants can show behaviors such as sensing, reacting to sensory inputs, and signaling in response to environmental influences, but there are stimulus-response explanations of what they were doing that do not require the attribution of conscious minds.

So why does Reber think that bacteria are conscious? He correctly notes that single-celled organisms have powerful ways of sensing their environments to detect sources of food and toxicity. Moreover, bacteria live in biofilms of large numbers of individuals that communicate with one another by secreting chemicals that spread important environmental information about food and toxins. Bacteria are capable of moving individually and collectively to get closer to food and farther from toxic substances. Perhaps sensing, reacting, communicating, and moving are best explained by the hypothesis that bacteria have some degree of consciousness.

But machines are also capable of sensing, reacting, communicating, and moving—for example, the self-driving cars that are being developed by Google, Uber, General Motors, and other companies. Reber thinks not only that such machines are not currently conscious but that they never could be, because he accepts the discredited thought experiment of John Searle that artificial intelligence is impossible because the symbols used by machines are inherently meaningless. I argued in chapter 3 that self-driving cars are capable of semantics in the same way as human brains through interacting with the world and learning about it. So machines that interact with the world can have meaningful representations even though they do not yet have consciousness.

Engineers know exactly how self-driving cars work because they built them and can explain their operations without invoking consciousness. Self-driving cars do not display behaviors such as pain, emotions, and imagery that consciousness helps to explain in birds and mammals. Self-driving cars and even thermostats refute Reber's claim that when an event is sensed, it is felt.

Another oddity of Reber's view is that he thinks that plants, which evolved from single-celled organisms, lack consciousness, even though they are capable of sensing, reacting, signaling other plants, and reorienting themselves toward the sun. An emerging enterprise called "plant neuroscience" tries to explain the behaviors of plants by analogy to how cognitive neuroscience explains the behavior of animals, although plants do not have neurons.

Reber's main reasons for attributing consciousness to single-celled organisms are not that doing so provides the best explanation of the available evidence but that this attribution solves philosophical problems. He thinks that his theory of the cellular basis of consciousness provides the most plausible answer to the problem of emergence. Consciousness is a property of objects unlike simple properties such as consisting of atoms and molecules or even firings of neurons, so hypotheses 2 through 9 face the problem of figuring out how consciousness became a property of wholes when it is not a property of their parts or a simple aggregate of the properties of their parts.

Fortunately, we have new theories of how consciousness could emerge as a property of large numbers of individual neurons, even though it is not a property of individual neurons. Stanislas Dehaene thinks that emergence comes from the broadcast of information across brain areas, whereas I argue in my book *Brain-Mind* that the key properties are patterns of firing of neurons, binding of these patterns into more complex patterns, and competition among the resulting patterns.

Both of these hypotheses about the emergence of consciousness in large brains have the advantage that they attribute consciousness to just those organisms for which we have evidence concerning pain, emotions, and imagery. We have no reason to attribute pain, emotions, or imagery to bacteria, so the attribution of consciousness is superfluous.

Another philosophical reason that Reber gives for his cellular basis of consciousness is that it provides a solution to the philosophical "hard problem" of consciousness: there is something that it is like to be conscious. But Reber's view does no better than others in accounting for the feeling aspects of consciousness, which can better be handled by breaking the problem down into specific aspects of pain and specific aspects of emotion. Without wallowing in the vagueness of "what it is like," specific aspects of conscious experience of emotion and imagery can be given neural explanations, as I show in *Brain-Mind.*

Therefore Reber's theory of the cellular basis of consciousness helps little with the philosophical problems of emergence and experience. Given his appreciation of scientific evidence, he should be able to recognize that the evidence for consciousness in single-celled organisms is much worse than the evidence for consciousness in self-driving cars, which already exhibit much more complex sensing, reacting, moving, and communicating than bacteria. Moreover, cognitive scientists are developing alternative hypotheses of how consciousness emerges through the complex operations of large brains capable of representing the world, learning about it, representing representations, and communicating with other brains.

Figure 6.3 diagrams the reasons for *not* attributing conscious minds to bacteria, and a similar analysis applies to plants. Bacteria lack the capacity of humans to report their thoughts and the ability to solve complex problems and learn to solve them better. The chief alternative to hypotheses that bacteria have conscious minds with some elements of intelligence is that they are much simpler systems whose cellular mechanisms enable them to sense, react, and communicate. Sensation is simpler than perception, which requires inferences about what objects are being perceived, not just chemical environments. Bacteria do not have feelings, according to the Attribution Procedure.

Lincoln Taiz and colleagues systematically debunk claims that plants possess consciousness and other mental properties. They deride the attribution of intelligence, cognition, problem solving, and learning to plants as based on misleading analogies, for example, between the chemical processes in

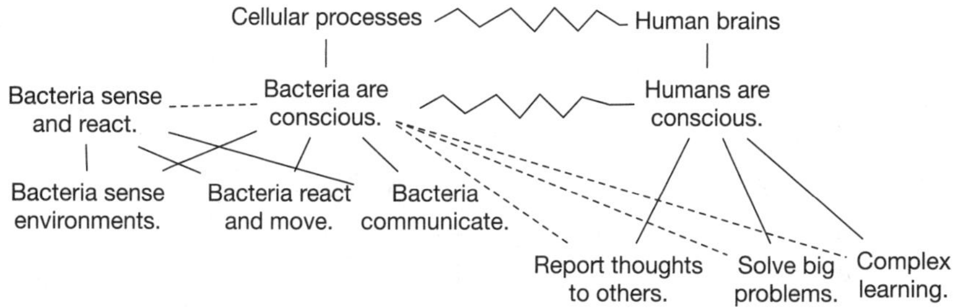

Figure 6.3
Diagram for attributing consciousness to bacteria. The solid lines indicate coherence resulting from explanations, and the dotted lines indicate incoherence. The jagged lines indicate incoherence resulting from disanalogy.

plants and the operations of neurons and neurotransmitters in animals. Taiz and his coauthors critique experiments intended to show associative learning by mimosa plants and argue that the sensitivity of some plant parts to anesthetics is best explained by chemical effects other than the reduction of pain. Unlike fish, discussed in the next section, plants have neither nociceptors (neurons that detect damage) nor opioid receptors that anesthetics can influence. Hence plants have neither the behaviors nor the mechanisms that would provide coherence to the hypotheses that they are intelligent and conscious.

To evaluate claims about intelligence in bacteria and plants, try giving them a report card like the ones I prepared for animals in report cards 4.1 and 4.2. Without metaphorical extrapolations, bacteria and plants get an F on all twenty features and mechanisms, despite their capacities for sensing, reacting, and signaling. Perhaps they could scrape by with a C for acting and communicating, but their performance on these features is far simpler than that of bees and the other animals I graded.

Do Fish Feel Pain?

Biologists are vigorously debating whether fish feel pain. Some argue that zebra fish and others feel pain because they display various behaviors that are best explained by the hypothesis that the fish are in pain. For example, fish whose flesh is damaged thrash around and lose their abilities to navigate, but they become calm and regain those abilities when given opioid painkillers. Fish have nociceptors (neurons that respond to potentially harmful stimuli) similar to pain-generating receptors found in mammals. Moreover, fish brains have opioid receptors similar to ones acted on by morphine in humans. The case for fish pain is thus a combination of saying that it explains behaviors and is itself explained by mechanisms similar to those that operate in mammals, as shown in figure 6.4.

Critics of fish pain challenge both the behavioral and the mechanism argument. They argue that thrashing is just an innate mechanism to avoid further damage, and they claim that just having nociceptors is no indication of pain, any more than just having light receptors is an indication of having visual experiences. Their contention is that fish brains, with only a few million neurons, do not have the capability of producing experiences of pain found in animals with much larger brains. This issue is important for practical as well as theoretical reasons because it has implications for whether fishing should be covered by regulations about animal welfare.

What is the burden of proof in such cases? Advocates of animal rights urge that lack of full scientific certainty should not be used as a reason for postponing measures to prevent the possibility of serious harm. Critics respond that there is no reason to favor fish over the welfare of humans who could benefit from the high-quality protein and fish oil that eating fish provides. Existing evidence does not show that fish have pain beyond a reasonable doubt, but that high standard also does not establish that fish do not have pain. So the appropriate standard is the preponderance of evidence concerning both behavior and mechanisms.

I am not convinced by the argument that fish brains are too unlike human brains to allow for the attribution of pain. As the discussion of ravens showed, animals lacking a prefrontal cortex, which contributes to

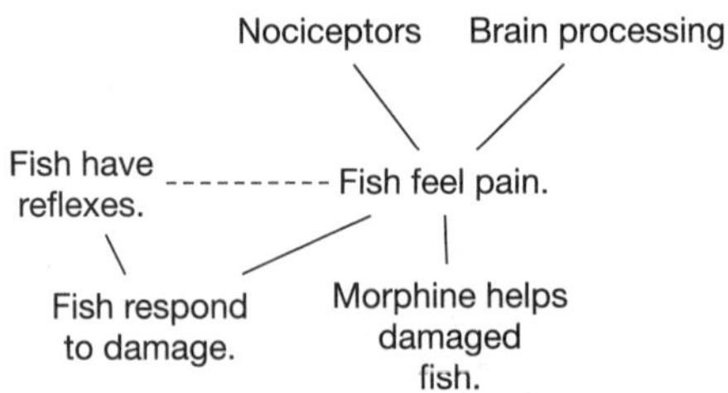

Figure 6.4
Diagram for attributing pain to fish. Lines have the same interpretation as in figures 6.1–6.3.

mammalian emotion, can still be judged to have emotions such as fear and anger. Fish have amygdala-like areas that receive bodily signals, so appraisal can still be going on in the fish pallium even without a ventromedial prefrontal cortex that contributes to the appraisal of such signals in mammals. Zebra fish have sleep cycles similar to mammals. What is less clear is whether the looping effects involved in human consciousness can operate in only a few million neurons.

We should distinguish a hierarchy of four levels of pain:

1. Damage detection. This detection could occur without any pain feelings, analogous to how a car signals that a door has been left open. Critics of fish pain think fish nociceptors signal only damage without processing it as pain.
2. Felt pain. Signals from nociceptors to the brain are processed neurally to produce the experience of pain.
3. Suffered pain. Nociception leads not only to pain experience but also to negative emotions such as distress, sorrow, or anger.
4. Reflective pain. Sometimes suffered pain goes beyond the moment in which it occurred to invade memory and imagination, leading to ongoing recollection of past pain and anticipation of future pain.

On the preponderance of evidence, I find it plausible that fish experience number 2, felt pain, but doubt that they have the neuronal capacity to experience suffered pain and reflective pain, which require appraisal with respect to goals and thinking across time.

Another controversy about fish and consciousness concerns their capacity for self-recognition in mirrors. There are reports that large-brained manta rays change their behavior in response to mirrors, and similar reports about a smaller species, cleaner wrasses. These studies are less definitive than ones where chimpanzees try to rub off the marks that they see on their own heads. Frans de Waal suggests that mirror recognition and self-consciousness may both be matters of degree rather than the all-or-none phenomenon usually assumed.

Are Cats and Dogs Jealous?

In chapter 1, I introduced the cats Zhanna and Pixie, who seems to get jealous when Laurette pets Zhanna. There have been no experimental studies of cats to discriminate between the jealousy and alternate hypotheses such as dominance, but new studies of dogs support the claim that dogs actually are jealous.

The general question is whether nonhuman animals have anything like human emotions. I expressed doubts about fish emotions, but mammals are much more similar to humans with respect to both behavior and brain mechanisms. As I did for pain, we can identify a hierarchy of emotions that vary across organisms: simple reward and threat responses, emotional feelings, mixed feelings, and feelings about feelings. To explore the issue of animal emotions, I present a debate in which a skeptic challenges the attribution of emotions beyond humans, and then resolve the debate using the example of animal grief. Then I examine the controversy about jealousy in dogs.

A Debate about Animal Emotions

Do nonhuman animals such as cats, dogs, and chimpanzees really have emotions like happiness, sadness, fear, and anger? Here is a dialogue between an advocate of animal emotions and a skeptic.

Advocate It is obvious that humans are not the only animals that have emotions. Anyone who has ever had a pet cat or dog knows that feeding them and petting them make them happy, whereas dangers make them afraid and angry.

Skeptic Not so fast. There is no doubt that such animals can be rewarded and threatened, but their behavior is no guarantee that they are experiencing the emotions that people have.

Advocate Your skepticism is bizarre. It reminds me of the philosophical problem of other minds, where the skeptic says, "I know that I have a mind, but how can I possibly know that anyone else has a mind?"

Skeptic The parallel between arguments about other human minds and ones about animal minds is not good because other people are much more similar to you than cats and dogs are. Can you provide a more substantial argument?

Advocate Gladly. The relevant argument pattern is what philosophers call inference to the best explanation, which is the standard way in science and everyday life of arguing about the existence of something you cannot directly observe. Most scientists believe in atoms because that hypothesis provides the best explanation of many phenomena in chemistry and physics. Similarly, we infer that the best explanation of other people's behavior is that they have minds just like us. Alternative explanations, such as those that suggest other people are robots controlled by space aliens, are utterly implausible. Analogously, the best explanation of the behavior of cats and dogs is that they are experiencing emotions.

Skeptic But wait, you neglect the fundamental principle of inference to the best explanation that you have to consider alternative hypotheses. For cats and dogs, we can explain their behaviors merely on the basis of reward mechanisms and threat response mechanisms that operate in all animals, including humans. When a cat is purring or a dog is wagging its tail, this response results from neural activity in its reward centers such as the nucleus accumbens. When a cat is yowling or a dog is growling, this results from neural activity in its threat detection centers such as the amygdala. These explanations are much simpler than making the additional assumption that cats and dogs are actually experiencing emotions of happiness and fear. Unlike people, pets cannot tell us that they are happy or anxious.

Advocate But thanks to neuroscience, we know that all mammal brains are similar with respect to the overall organization. In the argument about other human minds, not only do we use the hypothesis that other persons have minds to explain the behavior, but we know enough about human neuroanatomy to be able to explain that it is because they have brains like ours. We increasingly know the mechanisms by which brains make minds across humans and other mammals. Admittedly, this argument doesn't apply to insects, reptiles, and fish, which have much simpler brains. Whether it applies to birds is hard to say, because they don't have a prefrontal cortex, although they do share a similar brain structure—the nidopallium caudolaterale.

Skeptic The analogy between the brains of humans and nonhuman animals is not as good as you suppose. The brains of humans are far larger than those of cats and dogs, around eighty-six billion neurons as opposed to one or two billion. In particular, humans have a much larger prefrontal cortex, the area that is used for complex reasoning, so they are much more capable of making complex assessments of situations. If emotions were just physiological responses, then it would be plausible that animal emotions are the same as those in people. But physiology alone is not enough to discriminate between emotions such as fear and anger, which require an appraisal of situations with respect to situations and goals. This limitation is why nonhuman animals are incapable of complex human emotions such as shame, guilt, and fear of embarrassment.

Advocate We are talking not about emotions that depend on complexities of language and culture but about much more basic emotions such as happiness, sadness, fear, and anger. These do not require a linguistically and culturally mediated appraisal of the situation, merely that an animal can have some nonverbal ways of appreciating whether its goals such as food and safety are being satisfied or threatened. With respect to these, the neuroanatomy of mammals is sufficiently similar to that of humans to provide analogy-based support for the inference that animal emotions are the best explanation for their behavior.

Skeptic But the analogy remains weak, and you still haven't recognized that the alternative explanations of animal behavior based on reward and threat mechanisms are simpler than the attribution of emotions, making fewer assumptions about mental states. I suspect that your real reason for wanting to believe in animal emotions has nothing to do with inference to the best explanation. It's just a motivated inference: you want to believe that animals have emotions because you want them to feel about you the way that you feel about them. People love their cats and dogs, so they naturally want to be loved back.

Advocate Even if people have this motivation, it does not undermine the basic logic of the inference. Simplicity is not a stand-alone criterion for inference to the best explanation but has to be balanced against explanatory breadth. Attributing emotions to animals can explain aspects of their behavior that mere reward and threat mechanisms do not cover.

Skeptic To make this convincing, you need to specify the kinds of behavior that cannot easily be accounted for by reward and threat mechanisms, and to show that animal brains are capable of the appraisals that contribute to emotions in human brains. Until then, it is better to remain at least undecided about whether animals have emotions.

Grief

To break the deadlock in this debate, we only have to demonstrate one emotion in animals. A book by Barbara J. King convinced me that animals besides humans undergo grief, making it more plausible that they experience many other emotions.

In accord with the Attribution Procedure, we need to determine whether attributing grief to animals provides the best explanation of their behavior, taking into account alternative explanations of the relevant evidence, as well as deeper explanations of why animals feel grief. These deeper explanations use what is known about psychological and neural mechanisms to explain why animals grieve. Another contributor to the overall explanatory coherence of hypotheses that animals grieve is analogy between the explanations used for well-established human grief and animal grief.

According to Mark Bekoff, animals show universal signs of grief in responding to the death of a mate, family member, or friend. Like humans, grieving animals may withdraw and seek exclusion, sit motionless, lose interest in eating and sex, become obsessed with the dead individual, try to revive the individual, and remain with the carcass for days. These behaviors can all be explained by the hypothesis that the animals are grieving.

King provides abundant examples of grief behavior in animals across species that include cats, dogs, horses, rabbits, elephants, monkeys, chimpanzees, birds, dolphins, and whales. She also has examples of cross-species grief between pets and their owners, and even between a dog and an elephant.

For simpler emotions such as happiness and fear, there are alternative explanations of behavior, such as that a dog only seems happy because of a reward mechanism, or a cat only seems afraid because of threat reactions. But grief behavior does not seem to be explainable by such simple mechanisms, since it requires more complicated recognitions of attachment and loss.

What are the psychological mechanisms underlying grief? King thinks that humans and other animals grieve when they have loved. Animals who feel love for another go out of their way to be near to and interact with the loved one for reasons that go beyond mere survival. When animals are no longer able to spend time together because of death or separation, the loving animal will visibly suffer and act in ways that convey sadness.

We are beginning to understand the neural mechanisms for attachment and sadness. No single brain area is dedicated to love or sadness, because all emotions require the interactions of numerous brain areas such as the amygdala, orbitofrontal cortex, and striatum. All mammals have the same brain areas that underlie human emotions, while birds have different structures that seem to generate similar emotions. Mammals and birds all have neural mechanisms for representation, binding, competition, appraisal, and perception of bodily changes that are responsible for brains having emotions.

Figure 6.5 summarizes the reasons for concluding that elephants experience grief, including evidence such as staying with a dead body, explanation of grief based on attachment, competition with alternative hypotheses such as that elephants are just sad, and analogy with human explanations of grief behavior. More generally, I conclude that nonhuman animals do experience grief, because:

1. The hypothesis that animals grieve explains many cases of grief-like behavior.
2. There are no plausible alternative hypotheses that explain this behavior.
3. Known psychological and neural mechanisms operating in humans and other animals causally explain how grief comes about through attachment.

This kind of reasoning does not apply to all the emotions that have been assigned to animals. For example, in chapter 3 I described how the hypothesis that dogs experience guilt has been debunked by experiments finding that dogs show similar behaviors such as putting their paws over their heads regardless of whether they have actually done something wrong. So the alternative hypothesis of defensiveness is more plausible than assigning the complex emotion of guilt. Similarly, I see no reason to attribute to animals other complex states such as nested emotions (e.g., fear of embarrassment and longing for love). In contrast, the conclusion that animals experience grief by virtue of their psychological and neural mechanisms makes it plausible that they experience other simpler motions such as happiness, sadness, fear, anger, and surprise.

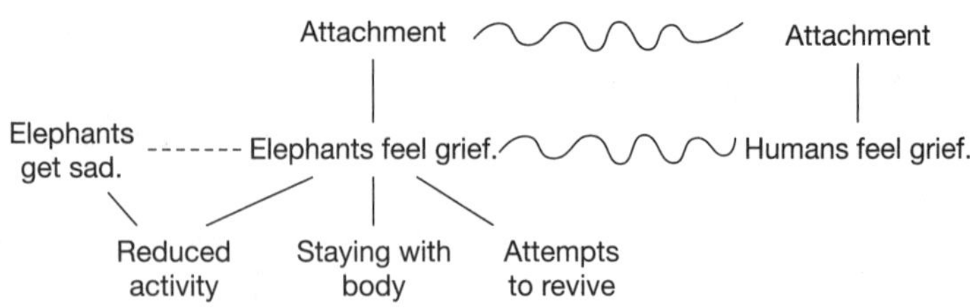

Figure 6.5
Diagram for attributing grief to elephants. Lines have the same interpretation as in figure 6.2, with the wavy line indicating analogical coherence. Evidence that humans feel grief is not shown but includes self-reports and many behaviors.

Jealousy

We can legitimately infer that cats or dogs are jealous if that hypothesis provides the best explanation of all the available evidence. Several factors contribute to determining the best explanation. First, how much does a hypothesis explain? For example, the hypothesis that Pixie is jealous of Zhanna explains why Pixie attacks Zhanna when Laurette pets her. Second, are there alternative hypotheses that could explain more? For example, maybe Pixie just wants to be dominant, although that does not explain why Pixie's attacks on Zhanna are more common when Zhanna gets her owner's attention.

Third, is the hypothesis in question simpler than alternative hypotheses in the sense of making fewer assumptions? An example of a nonsimple hypothesis is that Pixie attacks Zhanna because Pixie is controlled by space aliens, which requires additional assumptions about the existence and actions of the aliens. Fourth, does the hypothesis get additional support from an explanation of why it is true? Ideally, we could identify the psychological and neural mechanisms in Pixie that cause her to be jealous and attack Zhanna. Putting all these factors together, we might be able to accept the conclusion that Pixie is jealous because of its overall explanatory coherence.

Before determining whether cats and dogs are jealous, we should have some conception of what jealousy is. There is no standard definition of jealousy, but the concept can be characterized by the same method I used for intelligence and causality: standard examples, typical features, and explanations.

Familiar literary examples of jealousy include Shakespeare's *Othello* and Daphne du Maurier's *Rebecca*, and most adults can recall examples of jealousy from their own experience. The typical features of jealousy include the person who is jealous, the beloved who is the subject of jealousy, and the rival who is the object of jealousy. Othello is jealous because he thinks his wife Desdemona is romantically involved with Cassio. Jealousy differs from envy, which only requires two individuals: to be jealous also requires a threat to a relationship from a rival.

Additional typical features are emotions that go with jealousy, including fear of loss, threat to the relationship, sadness, anger, anxiety, and insecurity. The attribution of jealousy provides explanations of why people are experiencing these emotions and why they behave in ways ranging from withdrawal to aggression, sometimes even murder.

The first and weakest piece of evidence that cats and dogs are jealous is that most of their owners *think* that they are jealous. One study found that

81 percent of dog owners and 66 percent of cat owners report jealousy in their pets, which might be explained by supposing that the pets really are jealous. The alternative hypothesis is that people who are attached to their pets exaggerate their mental complexity. For example, 74 percent of people report that their dogs sometimes feel guilty, but experiments showed that dogs behave the same way when they did nothing wrong. Many religious pet owners think that their dogs and cats have souls, for which no evidence exists at all.

Much stronger support for jealousy in dogs comes from a study by Christine Harris and Caroline Prouvost. They adapted an experimental design used to identify a nonverbal form of jealousy in infants as young as six months. Infants had more negative reactions when their mothers paid attention to another infant than when the mothers paid attention to a book. Analogously, when dogs saw their owners interacting with a fake but realistic dog, the dogs displayed more aggression in the form of biting and snapping, plus more attention seeking and disruption of interaction. In contrast, when the owners paid attention to a jack-o'-lantern or a book, the dogs did not display more aggression and attention seeking. A skeptic might worry that the dogs were simply reacting to a strange object, the fake dog, but the dogs in the experiment did not seem to realize that the dogs were fake and sniffed their rear ends.

Nevertheless, it is possible that the dogs' behavior resulted from mental states other than jealousy, for example, from a desire to establish dominance over the new dog or a resentment that the fake dog was challenging its territoriality. A study by Judit Abdai and others changed the experiment to reduce alternative interpretations. The rival dogs toward which the owners showed affection were real rather than fake, ruling out the hypothesis that the dogs' aggression was simply caused by a strange object. To avoid the dogs being territorial, the experiments were performed in an unfamiliar place rather than in the dogs' homes. The experiments also compared the behavior of dogs with respect to both familiar and unfamiliar rivals to control for issues of dominance by eliminating an association with rank. Other experimental arrangements made it implausible that the dogs' responses resulted from protectiveness, playfulness, or boredom.

The experimenters concluded that dogs display behaviors that have been referred to as jealousy, similar to those observed in children under two years of age. Nevertheless, the authors resist the conclusion that the dogs in their experiments are actually experiencing the emotion of jealousy, preferring the more cautious conclusion that the dogs are exhibiting jealous behavior.

What are the psychological and neural mechanisms that make dogs jealous? The hypothesis that dogs are jealous would be deepened if we could explain why dogs are jealous in terms of psychological and neural mechanisms. The relevant psychological mechanisms are attachment and fear of loss. We have ample evidence that pets become emotionally attached to their owners, as seen in the distress they exhibit when owners leave them alone, and the grief behavior that is sometimes observed when owners die. Then the attachment that pets feel for their owners explains why they are threatened by their owners' attention to other animals. Owners of pets report that the degree of jealousy they observe in cats and dogs varies with how much the cats and dogs like their owners. This observation might also explain why people think dogs are more jealous than cats, since generally dogs are more attached to their owners than cats are.

Psychological mechanisms are increasingly being understood in terms of underlying neural mechanisms. It would be folly to look for a jealousy center in the brains of humans and other animals, because emotions and other kinds of cognition involve interactions among many brain areas. Nevertheless, researchers are using brain scans such as functional magnetic resonance imaging (fMRI) to identify brain areas that interact to produce mental states such as emotions.

Recently, experimenters have trained dogs to stay still in fMRI machines that identify brain areas active during presented tasks. An article by Peter Cook and his colleagues described a brain imaging experiment that they think provides evidence for jealousy in dogs. They scanned the brains of dogs who watched their caregivers give a food reward to a fake dog, and predicted that the dogs would display greater activation in the amygdala than when their caregivers merely put the food in a bucket. Cook and his colleagues also predicted that dogs who had previously been judged to be more aggressive would show more amygdala activation than ones with less aggressive temperaments. Both predictions were confirmed.

By itself, this experiment does not demonstrate that a scanned dog is jealous of the fake dog, because alternative hypotheses exist. The dog may be reacting merely with a sense of annoyance, envy, hostility, or unfairness rather than jealousy. Activation of the amygdala does not pinpoint jealousy, as the amygdala is also associated with other emotions such as anxiety, anger, fear, and even some positive experiences.

Nevertheless, this experiment is relevant to whether dogs are jealous because it points to a neural mechanism that can explain how dogs become jealous. When a dog sees its owner being generous to another dog, then it responds to the situation by the firing of neurons in a part of the brain that is associated with negative emotions and aggression. Then amygdala activation leads to aggressive action such as growling and biting. Much further research in both humans and pets is needed to provide more details about the neural mechanisms of jealousy, but the Cook study is a start.

Do the Harris, Abdai, and Cook studies together suffice to show that dogs are jealous? There are no dire ethical or practical consequences of accepting the conclusion either that dogs are jealous or that dogs are not jealous, so the burden of proof should be preponderance of evidence rather than reasonable doubt, where preponderance can be assessed by explanatory coherence.

My current assessment is that reasonable doubt persists about whether dogs are jealous. It is possible that their observed behaviors in ordinary life and in scientific experiments result from reactions that do not involve the perception of rivals as relationship threats or the combination of emotions such as fear, anger, and sadness that add up to jealousy. Nevertheless, I think that the preponderance of evidence drawn from behavioral experiments, brain imaging, and owners' observations supports the conclusion that dogs are indeed jealous. Figure 6.6 displays how this hypothesis provides a more coherent explanation of all these pieces of evidence than apparently simpler explanations that avoid attributions of jealousy.

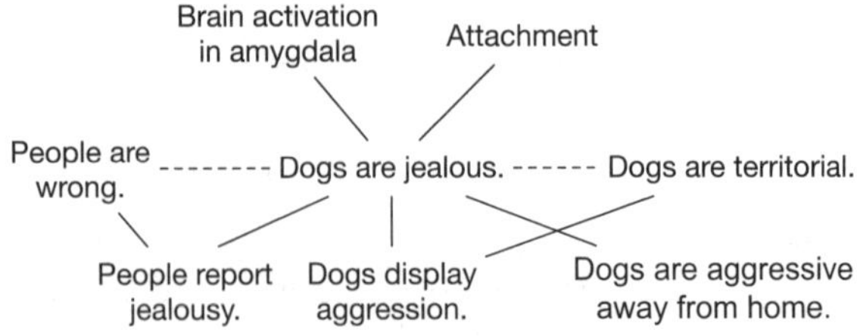

Figure 6.6
Diagram for attributing jealousy to dogs. Lines have the same interpretation as in earlier figures. The analogy to human jealousy is not shown.

The conclusion about dog jealousy has problems concerning the complexity of animal cognition and the vexed issue of consciousness. The attribution of jealousy to animals other than humans can be challenged by arguing that animals do not have the cognitive complexity to be jealous. In adult humans, jealousy requires a judgment such as "my relationship to my beloved is threatened by the beloved's interest in a rival." Cats and dogs do not even have a full sense of self, as shown by their inability to recognize themselves in mirrors. So they do not even have enough of the concept *I*, let alone the sense of a relationship that involves a triad of three individuals: the lover, beloved, and rival.

However, infant studies suggest that jealousy can arise long before children start to recognize themselves in mirrors around the age of eighteen months. Being jealous does not require having a full sense of self, only a minimal awareness of being distinct from others and being in some emotional relationship. Like infants, cats and dogs are capable of such awareness, as shown by the ways in which they use and touch their own bodies, for example, when cats lick themselves to become clean. Pets may not be able to verbally describe their relationships with their owners, but many behaviors suggest the existence of an affectionate bond. A challenge to this bond in the form of perception of the owner's interest in another pet is enough to prompt an emotion that is at least an approximation of jealousy in humans.

Similar reasoning justifies the attribution of conscious experience of jealousy in pets, not just jealous behavior. When people attribute consciousness to themselves, the explanatory coherence of the conclusion is strong because we each have a range of conscious experiences, including pain, emotions, thoughts, and self-awareness. Despite the protests of behaviorists and other skeptics about consciousness, there is no good alternative explanation of your experiences and behaviors than that you are actually conscious. As I argued earlier, the similarities between others people's behaviors and yours, and between their measurable brain processes and your own, make it plausible that other people are conscious too.

The analogy gets weaker when we try to extend consciousness to infants, but we know that their brains have emotional structures and functions much like those of adults. We have similar knowledge about other mammals such as cats and dogs, all of which have brain areas such as the amygdala and cortex that contribute to human emotions, even if human prefrontal areas are larger. Some reasonable doubt may remain that cats, dogs, and infants are not really conscious, but the preponderance of evidence suggests that they are. So the conclusion that pets experience jealousy is not ruled out by problems about consciousness or about complexity of cognition.

The Harris, Abdai, and Cook experiments only address the question of whether dogs are jealous, leaving open whether cats are jealous too. Consistent with their larger body size, dogs have brains with around twice as many neurons as cats. But the brain organization of cats is the same as that of dogs and other mammals, so brain structure does not lead us to expect any difference with respect to emotions.

On the other hand, through the approximately fifteen thousand years that dogs have evolved relationships with people, dogs have acquired cognitive and emotional abilities that are not found in cats. Dogs are generally more attentive and attached to their owners than cats, so it is possible dogs are more prone to jealousy. Moreover, both dogs and cats may display differences across breeds with respect to degrees of attachment and aggression that are connected to jealousy. These differences, along with the lack of any behavioral and neural experiments that support the existence of jealousy in cats, suggest more caution about identifying jealousy in cats than makes sense for dogs.

What about Pixie and Zhanna? Pixie is a Burmese, a breed well-known for its doglike affection for people. In contrast, Zhanna is a British shorthair, a breed much better known for detachment than for physical affection. Perhaps this explains why Pixie is much more aggressive in reacting against affection to Zhanna than vice versa. Neither seems particularly bothered when I am affectionate with the other one, presumably because I am much less important to them than their owner, Laurette. In contrast to the growing scientific evidence that dogs are jealous, I only have a personal hunch that Pixie is also jealous.

Can Apes and Other Animals Think with Analogies?

In chapter 2, I presented analogical thinking as one of the mechanisms of human thinking that is responsible for many cases of flexible problem solving and creativity in science and other fields. In 1995, Keith Holyoak and I published a book on analogy called *Mental Leaps*, which included a chapter written by Keith titled "The Analogical Ape." But by 2008 Keith had changed his mind and claimed with Derek Penn and Daniel Povinelli that only humans can reason with analogies. In contrast, some researchers have attributed analogical thinking to chimpanzees and other animals. When I began looking at this question, I was neutral but examined the studies that claim to show that nonhumans can perform analogies. My conclusion is that animal performance lacks important aspects of thinking used by people in analogical arguments, proportional analogies, and analogical problem solving.

Animals Recognize Similar Relations

Research on learning in animals has long showed their ability to recognize similar visual features such as shapes, colors, and patterns. For example, if a pigeon is presented with a red square and then with two shapes including a red square, it is rewarded if it pecks at the red square. Through reinforcement learning, many animals can learn to identify similar features in such tasks.

In 1981, research with a chimpanzee extended judgments of similarity to apply to relations as well as similar features. Sarah the chimpanzee had been given extensive language training and acquired the ability to use tokens to discriminate between *same* and *different*. For example, she could present one token to indicate that a red square is similar to another red square, but the other token to indicate that a red square is different from a blue circle.

Sarah also learned to recognize similar and different relations. Figure 6.7 shows a relational match with an initial stimulus followed by a choice of two potential matches. On the left, the initial two squares are the same shape, so Sarah had to recognize a match with choice 1 but not choice 2. In contrast, on the right, the initial shapes are different, so Sarah had to recognize a match with choice 2 but not choice 1. Sarah learned to make such discriminations with a variety of relations such as part-whole.

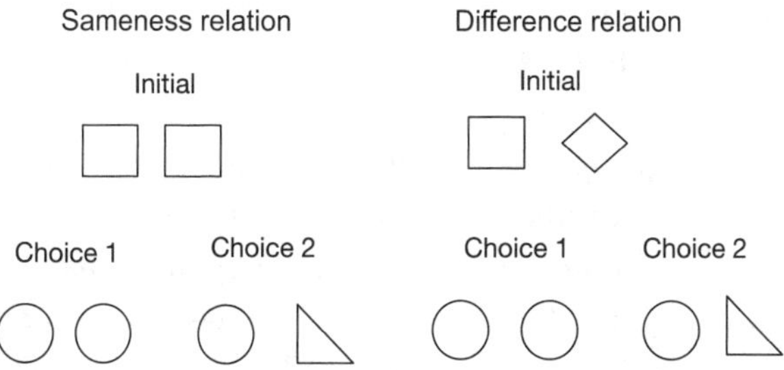

Figure 6.7
Relational similarity task in which choice 1 is correct on the left side, and choice 2 is correct on the right side.

Psychologists first thought that Sarah was special in her ability to identify similar and different relations because of her unusual linguistic training, but later research found that other chimpanzees, baboons, crows, parrots, and pigeons could also perform tasks like the one shown in figure 6.7. These tasks are analogy-like in that they require noticing relations among relations. You could say that on the left side of figure 6.7, choice 1 is analogous to the initial pattern, whereas on the right side choice 2 is analogous to the initial pattern. But this task is much simpler than the analogies that people think with.

Analogical Arguments

Long before analogy became a hot topic in cognitive science, it was important in philosophy going back to Aristotle. In the most influential logic text of the twentieth century, Irving Copi said that "every analogical inference proceeds from the similarity of two or more things in one or more respects to the similarity of those things in some further respect." For example, Copi says, you might expect to enjoy a book by a familiar author if you have read and enjoyed other books by the author. This characterization applies to the recommender AI systems discussed in chapter 3.

Copi emphasizes that analogical arguments can be weak but can be appraised by assessing (1) the number of entities between which the analogies are said to hold, (2) the number of respects in which the things involved are said to be analogous, (3) the number of disanalogies or points of difference, and (4) the extent to which the similar respects are causally relevant to the conclusion. For example, your judgment that you will like another book by the same author is stronger if (1) you have read more books by that author, (2) the new book you are considering is similar in many respects to the old one, (3) there are few points of difference between the new and old books, and (4) the points of similarity are causally relevant to your enjoying books, for example, the topic and style that give you pleasure.

As a logician, Copi assumed that analogical inference is linguistic, but his criteria apply equally well if the mental representation of the instances and similarities is nonverbal through modalities such as seeing, hearing, touching, tasting, and smelling. For example, I can infer that a kind of blue cheese I have never tried (Spanish Cabrales) will taste good because it looks similar to ones I like to eat such as French Roquefort and Italian Gorgonzola. My inference is strengthened if I know that the cheese's blue veins are made by the mold penicillium, which causes their unusual taste.

Copi's analogical arguments are more complicated than the tasks performed by animals, because they require inferences that go beyond the mere recognition of relations among relations. For example, the book analogy extrapolates from the properties of the familiar books to the projected property of the new book, ideally extending the causal relations that hold between reading the old books and enjoyment to a corresponding relation for the new book. The mere recognition of relational similarity or difference does not produce such projections.

Proportional Analogies

Inferences that go beyond recognizing relations among relations are also required for the proportional analogies often used in IQ tests, with the structure "*A* is to *B* as *C* is to what?" People can readily answer questions such as "foot is to shoe as hand is to what?" Unlike the arguments discussed by Copi, the answer "glove" has to be generated, first by figuring out that the relation between foot and shoe is that shoes are worn on feet, and second by transferring the relation "worn on" to "hand" to come up with the answer "glove." This transfer from feet to hands requires an inference that goes beyond information that is present in the initial information, not just recognition of similarities among relations. So, like Copi's arguments, proportional analogies are more inferential than relation matching by animals.

Analogical Problem Solving

Analogies contribute much more to human thinking than arguments and test questions, being partly responsible for many of the greatest accomplishments in science (e.g., Darwin's theory of evolution), technology (e.g., the stethoscope), and literature (e.g., autobiographical novels). Analogical problem solving typically starts with a target problem to be solved, such as figuring out how to throw a party. It then retrieves a similar source problem, such as remembering a previous successful party. This source can be mapped to the target to suggest an approximate solution to the target, for example, including the same kinds and amounts of alcohol. The source problem is adapted to provide a solution to the target problem by transferring causally relevant features from the source to the target, for example, by appreciating that having abundant beer and wine makes a party more fun.

Table 6.1 shows that this account of analogical problem solving encompasses relational similarity, analogical arguments, and proportional analogies.

Table 6.1
How relation matching, analogical argument, and proportional analogy are all kinds of analogical problem solving

Type	Source problem	Target problem	Mapping	Transfer
Relation matching	Initial pattern	Choice pattern	Correspondences between relations	None
Analogical argument	Familiar thing	New thing	Old thing and respects to new thing and respects	Property of new thing
Proportional analogy	*A* : *B*	*C* : ?	*A* to *B*, and *C* to ?	Source relation to target relation and answer

The major difference is that relation matching does not add any new information to the target problem: no new inference occurs beyond noticing the connection between the source initial pattern and the target choice pattern. Similarly, analogies in science, technology, art, and everyday life all require adapting a source to transfer solutions to a new problem, for example, when Tolstoy drew on his own life to generate the fictional world of *War and Peace*. Another difference between animal relational matching and complex analogical problem solving is that the latter can involve layers of causal relations. For example, Darwin argued that natural selection (caused by competition to survive and reproduce) causes evolution of species just as artificial selection by farmers causes development of new breeds.

Because relation matching by animals lacks inferential transfer and relational hierarchies, I am reluctant to attribute analogical thinking to chimpanzees, crows, and others that can do the task shown in figure 6.7. It was generous to give ravens and chimpanzees a C for analogy on the mechanism report card 4.2, but their ability to map relations to relations suggests that they deserve something better than an F. One advantage of the report card format is recognizing that mental features and mechanisms are often a matter of degree rather than strictly yes or no. Perhaps expanding my grading scheme to allow a D for analogy by chimpanzees and corvids would be appropriate.

In chapter 4, I raised the question of whether imitation learning is a kind of analogical thinking. For example, when a chimpanzee learns to crack a nut by copying another chimp, it might be like an analogy where the target problem is how to get the nut open and the source problem is the other chimpanzee using a rock as a tool. Unlike the relational matching task, this really would be a case of analogical transfer through the realization of "I could use a rock too." However, given that it takes years for chimpanzees and capuchin monkeys to learn to crack nuts, such imitation sounds more like repeated trial-and-error learning spurred by observing the other rather than analogy that works quickly by mapping and transfer.

Even when an animal or person learns quickly by imitating another, the process is simpler than analogical mapping and inference. When I learned how to do tai chi, I copied the movements of the instructor one by one, for example, moving my hands from my sides up over my head while breathing in, and then moving them down in front while breathing out. Learning this exercise required only sequential copying, not mapping a whole causal structure and using it to make a new inference. Similarly, the dolphins mentioned in chapter 4 that learned to tail walk from others only needed to repeat sequential bits of their routine, not map the whole structure.

Will Computers Have Minds?

In chapter 4, I established that machines already display many features and mechanisms of intelligence to some extent, but we naturally want to know whether AI will catch up to human minds with respect to imagery, consciousness, emotions, and creativity. Deciding that machines can think is more complicated than just having them pass the Turing test of imitating humans. Imitating a human is not sufficient for having a mind, because it can be accomplished by linguistic tricks, as in chatbots like the mechanical therapist Woebot. Imitation is not even necessary for having a mind, because a machine might be tripped up in answering some trivial question such as "can you touch your right elbow with your right thumb?" even though it might be overall much smarter than the human. Instead, in line with the Attribution Procedure, we can use what we now know about human features and mechanisms to extrapolate about future machines.

Will Computers Be Conscious?

Max Tegmark thinks that intelligence is "substrate independent," which means that it can potentially operate in different kinds of hardware, including brains, digital computers, and many other machines. In contrast, Todd Feinberg and Jon Mallatt maintain that consciousness evolved from inherently biological processes that depend on the nature of living organisms on this planet. Compare the difference between sexual reproduction and arithmetic: sex depends on particular biological entities such as ova, sperm, and body parts, which make it substrate dependent; but arithmetic can be performed by many different devices, including minds and computers. Is consciousness like sex or like arithmetic? What are the prospects for the successors of IBM Watson, AlphaZero, driverless cars, and other machines eventually developing consciousness?

Animals do not provide a good guide to the development of machine consciousness because their physical structures and brains are so different from the machinery used in computers and robots. A simple kind of animal consciousness is pain, which may even operate in fish with only a few million neurons. But why would anyone ever try to build a robot that experiences pain? Nociceptors are a biological means of detecting damage to cells, but machines do not have cells, so there is no reason for them to have nociceptors. We have other ways of building damage detectors into machines, for example, by having electronic sensors that can tell when a bumper is dented. Driverless cars and other robots can acquire mechanisms for attention, evaluation, and motivation that operate independently of the internal sensors and emotions that perform these functions in humans.

Chapter 3 described how self-driving cars are already using combinations of sophisticated sensors based on cameras, radar, lidar, and GPS. Such cars already have perceptions, but what would it take to give them perceptual consciousness? One recursive property that would help would be the ability not only to make perceptual judgments such as that an object is a human body but also to have a representation of this perception, for example, as part of an ethical judgment that running over a human body is undesirable. Self-driving cars do not yet have this reflective capability, but combining them with the greater linguistic resources of a system such as IBM Watson might provide it. I already suggested that Watson needs the perceptual capabilities of driverless cars to enrich its semantics, and driverless cars similarly need the linguistic abilities of Watson to enhance their potential for consciousness.

Computers already have a limited capacity for self-representation, for example, when they tell you what operating system they are running and how much storage they are using. But to acquire consciousness, computers are going to need richer processes, such as the modal retention that operates in semantic pointers to retain approximations to sensory and motor inputs.

Simulations of semantic pointers already run on computers, including ones that use neuromorphic chips that simulate millions of neurons and interface with cameras and robotic arms. So I think that these are the likeliest candidates for producing something like consciousness in future machines. Another mechanism relevant to consciousness is competition, in which different representations vie to determine which can be important enough to break through into consciousness. Such competition is easy to implement in computer systems. Another neural mechanism that may be important for human consciousness is the ability, described by Stanislas Dehaene, of brains to broadcast information from some local parts to others, which again should not be difficult to implement in machines.

It would be rash to suppose that machines could *never* be conscious, although it is hard to see why future engineers and scientists would go to the effort to produce conscious machines. Another possibility is that future robots will evolve by building themselves, just as some computer programs today can modify themselves. In this scenario, machines will not be programmed to be conscious but will develop consciousness on their own in ways not currently predictable. In chapter 7, I discuss whether having consciousness in machines is morally desirable.

Will Computers Have Emotions?

What would it take to make a robot emotional, and would we ever want robots to have that capacity? According to obsolete ideas, rationality and emotion are fundamentally opposed because rationality is a cold, calculating practice using deductive logic, probabilities, and utilities. But abundant evidence from psychology, neuroscience, and behavioral economics shows that cognition and emotion are intertwined in the human mind and brain. Although we can point to cases where emotions make people irrational, for example, when a person loves an abusive spouse, in many other cases good decisions depend on our emotional reactions to situations. Emotions help people to decide what is important and to integrate complex information into crucial decisions, so it might be useful to try to make a robot that has emotions too.

Another reason for wanting emotional robots is the prospect that they will be used to look after human beings, as is increasingly common with old people in Japan. Having robots with emotions might make them better at understanding and caring for people.

Moreover, as robots become more capable of autonomous actions, we have a greater need to ensure that they act ethically. We want robots on highways and battlefields to act in the interests of human beings, just as good people do. Ethics is not just a matter of cold calculation but also requires emotional processes such as caring, sympathy, and empathy. The emotional makeup of human brains makes us capable of caring about other people and understanding them empathically. So if robots are going to be ethical in the way that people are, they need emotions.

Estimating the feasibility of making robots emotional depends on understanding what makes *people* emotional. Today we have three main theories about human emotions, based on appraisal, physiology, and social construction. The cognitive appraisal theory says that emotions are judgments about the relevance of the current situation to a person's goals. For example, if someone gives you $1 million, then you will probably be happy, because the money can help you to satisfy your goals of surviving, having fun, and looking after your family. Robots are already capable of doing at least a version of appraisal, for example, when a driverless car calculates the best way of getting from its current location to where it is supposed to be. If emotions were just appraisals, then robot emotions would be imminent.

However, human emotions also depend on physiology. Responses such as being happy to get a pile of money are tied in with physiological changes such as heartbeat, breathing rate, and levels of hormones such as cortisol. Because robots are made of metal and plastic, it is unlikely that they will ever have the inputs from bodies that help to determine the experiences that people have, the feelings that are much more than mere judgments. On the theory that emotions are physiological perceptions, robots will probably never have human emotions, because robots will never have human bodies. It might be possible to simulate physiological inputs, but the complexity of the signals that people get from all their organs makes this unlikely. For example, the digestive tract contains 100 million neurons that send signals via the vagus nerve to the brain based on the activities of billions of stomach cells and bacteria.

The third prevalent theory of emotions is that they are social constructions, dependent on language and other cultural institutions. For example, when $1 million falls into your hands, your response will depend on the language with which you describe your windfall and the expectations of the culture in which you operate. If robots ever get good at language and form complex relationships with other robots and humans, then they might have emotions influenced by culture.

I think that these three theories of emotions are complementary rather than conflicting, and the semantic pointer theory of emotions shows how to combine them in brain mechanisms. So maybe robots could get some approximation of human emotions through a combination of appraisals with respect to goals, rough physiological approximations, and linguistic and cultural sophistication, all bound together in semantic pointers. Then robots would not get human emotions exactly, but some approximation could perform the contributions of emotions for humans listed in chapter 2: evaluation, relevance, attention, motivation, memory, and communication.

Computers are already able to simulate some of the aspects of emotions, such as appraisal, but no one claims that computers are actually experiencing emotions. The physiological side of human emotions depends on dozens of streams of information coming from different bodily organs. Duplicating the streams would require replicating the outputs of all the organs, as well as the layers of interpretation of this information taking place in various brain areas such as the amygdala, insula, and orbitofrontal cortex. Such duplication may be physically possible, but it is so technologically challenging that no robot builder will ever be inclined to do it: there are other less biological, more electronic ways of accomplishing the contributions of emotions. So emotional consciousness will likely be missing in the intelligent machines of the future, along with connected mental states such as empathy and sympathy that are important for human relationships. Emotions appear to be more like sex than like arithmetic, dependent on our embodied substrate.

Will Computers Be Creative?

In chapter 3, I gave some modest examples of creativity produced by artificial intelligence programs. IBM Watson has a chef version that has generated interesting recipes. AlphaGo has come up with moves that Go experts view as surprising and effective. Deep learning has even been used to generate sonnets in the style of Shakespeare.

Marcus du Sautoy describes recent impressive examples of computer creativity in art and music. A Paris collective produced attractive portraits using generative adversarial neural networks that combine a trained network to generate possible works and another to evaluate them. They trained their program on fifteen thousand portraits and used it to generate depictions of a whole fictitious family. One of these portraits sold at auction for more than $400,000! Du Sautoy also describes a program called DeepBach, which was trained on Bach chorales and generated new compositions that 50 percent of humans judged to have been composed by Bach himself. Computers have also been used to generate novel mathematical proofs. Artificial intelligence is already producing results that fit with the typical conditions of creativity by being new, surprising, and valuable.

Nevertheless, Sean Dorrance Kelly insists that the human achievement of creative innovation will never be surpassed by machines. A Harvard philosopher writing in the *MIT Technology Review*, Kelly gives several reasons for the inherent superiority of humans. He says that human creativity is socially embedded, depending on norms that govern our culture. He argues that a machine could never bring about major changes in music such as Schoenberg's dissonant compositions. Current programs merely mimic musical styles but could not invent a whole new way of making music. Computers can apply preexisting standards but do not have a vision of human good that could be used to change standards for judging the quality of music.

Kelly rejects claims for creativity for game-playing program such as AlphaGo because they are constrained to narrow domains. Human creativity is more general, following norms for how humans act, but also having the ability to change those norms. Creative mathematicians are not just theorem provers but come up with new and unexpected forms of reasoning recognized by other mathematicians as significant. Kelly concludes that AI can be a tool for enhancing human discovery but can never be an autonomous creative agent.

Kelly is right that AI creativity is not yet close to that of great composers or mathematicians, but he underestimates the possibility of future progress. In my book *Brain-Mind*, I show how some of the highest achievements of human creativity involve the discovery of new methods. For example, Galileo did not merely discover the moons of Jupiter; his new method of using a telescope to explore the heavens led to many other discoveries. No AI program today can generate new methods, but I described two ways of doing so by generalization and by analogy.

Method generalization can produce a new rule from only one example:

Input: One or more goals, techniques consisting of one or more steps, and a problem solution showing that using the steps leads to accomplishment of the goal.

Output: A method with the structure "If you want to accomplish the goals, then use the technique consisting of the steps."

Process: Identify the steps that led to the goal, and generalize them into the method.

For example, Galileo needed only one case of using a telescope to look at Jupiter to produce a new, surprising, and valuable way of investigating the planets. Schoenberg introduced dissonances in a ballad he wrote in 1907 and then generalized the absence of harmonics into the method of atonal music.

Method analogizing uses two or more cases to produce a new method:

Input: A rule (method) that operates in one domain, and another similar domain.

Output: A new rule that provides a method for operating in the new domain.

Process: Analogically adapt the original rule to provide a method for the new domain.

For example, Mark Zuckerberg developed Facebook as a new method for social interaction by combining the idea of books previously used for student information with web-based computational methods.

Because generalization and analogy have been extensively studied in AI, and because rules are a familiar mental and computational representation, it should not be hard to implement the two ways of producing new methods in some future system of machine intelligence. With rich domain knowledge, such a system could then be capable of generating new ways of doing things.

But would the system be capable of changing norms, which Kelly insists is a mark of full human creativity? Kelly does not say what social norms are, but I give a psychological theory of how they work in my book *Mind-Society*. Norms can be understood as mental rules with an if-then structure, where the if and then parts can be nonverbal and emotional. For example, the Western social norm of shaking hands when you meet someone can be expressed as *if meeting, then shake hands*, where shaking hands is a physical action. This rule is associated with positive emotions such as the pleasure of meeting someone, and with negative emotions such as disapproval of anyone violating the norm. Implementing such norms fully in machines would require that they have emotions, but an approximation could be accomplished by if-then rules.

Emotions are generally an important part of creativity because a creative product is supposed to be surprising and valuable. Assessments of value and surprise are emotional reactions, and I just described the difficulties of getting computers to have emotions. Does this mean that computers could not be creative? No, because computational ways of assessing surprise and value do not need to use the emotional means that work well in human bodies. The computer may not be capable of being surprised, but it could at least carry out the cognition of determining that an event such as observing moons near Jupiter is contrary to previous expectations. A new poem generated by a computer may not stem from the emotional experience of a human, but it may still be valuable in its ability to generate emotional responses in the people who read it.

Therefore it is premature to rule out machines becoming as creative as people, even though achieving such a goal will require major advances in psychology, neuroscience, and artificial intelligence. Merely making computers faster will not suffice, and creativity will not be solved by familiar techniques such as deep reinforcement learning. Fully creative computers are not just around the corner, but they may well be part of future technological progress, unless people decide on ethical grounds not to make computers more creative than they already are.

Minds Began Gradually

According to theological views, minds began suddenly when God created them. The biological story is much more complicated, because minds combine at least twenty features and mechanisms, each with gradations such as different kinds of learning and different aspects of analogy use. Even bacteria do a lot with only a single cell by sensing, reacting, moving, and signaling other bacteria.

Nevertheless, bacteria and plants are extremely limited in features such as problem solving and learning and in mechanisms such as consciousness, so I see no reason to attribute minds and intelligence to them. But fish, insects, and cephalopods such as octopuses seem to have the neural capacity to carry out many features of intelligence. They may even have some capacity for emotions, although a broad range of emotional states and reactions probably began with mammals and birds with much larger brains.

The examples of tool use by nonhuman animals such as chimpanzees and ravens show that they have some understanding of causality, for example, that hitting a nut with a rock causes the nut to break open, and using a beak to put pressure on a stick causes the stick to bend. But these are merely causal relations between observable things, not the postulation of unobservable causal mechanisms that operate in human enterprises as diverse as science and religion.

Accordingly, we can conclude that mind developed gradually in stages like the following:

1. Elementary sensing, acting, and signaling, as in bacteria and plants.
2. Some problem solving and learning, possibly with conscious experience, as in insects and fish.
3. Much richer problem solving and learning including emotions and consciousness, as in birds and mammals.
4. Full intelligence including analogy and hidden causes, as in humans.

With respect to the attribution of mind, I think that current evidence warrants being a killjoy about nonliving things, plants, and bacteria. But a reasonable romanticism is appropriate for birds and mammals.

From this perspective, we can ask whether intelligent life is likely to have evolved on other planets. A calculation by Tom Westby and Christopher Conselice estimates that at least thirty-six other civilizations exist in our galaxy, but these authors assume that intelligent life will develop elsewhere as it did on Earth. However, some of the steps that led to intelligence on our planet were highly improbable, given that they seem to have occurred only once and depend on mechanisms that would produce them rarely. These steps include the evolution of eukaryotes that have energy advantages over cells without nuclei, the sudden extinction of dinosaurs that allowed mammals to become more prominent, and whatever genetic mutations enabled brains to function recursively. In contrast, other steps toward intelligent life probably occurred repeatedly, such as the formation of amino acids, the development of cells, the evolution of vertebrates, and the inventions of cooking and agriculture. Overall, however, my guess is that human intelligence is a cosmic fluke. Let's not blow it.

What about computers, robots, and other machines? In chapter 3, I argued that current machines have some features and mechanisms of intelligence but lack emotions and consciousness. Artificial intelligence is rapidly evolving, but extension to minds that include imagery, consciousness, and emotions is far away. In the next chapter, I argue that such distance is worth maintaining.

7 The Morality of Bots and Beasts

How should we treat machines and animals? They are not as smart as human beings, but they still have many of the features and mechanisms that make people intelligent. So what moral obligations do we have toward them? Intelligence is not the only factor that goes into moral obligations, as newborn babies have minimal intelligence but deserve full moral consideration and protection.

The ethics of machines and animals are not just abstract philosophical discussions, because many practical questions increasingly need to be answered. With respect to machines, the biggest question is whether actions should be taken to prevent the robot apocalypse—the complete destruction or domination of humans by computational intelligence. Because of the limitations of AI today, this issue is not pressing, but many other ethical questions about machine intelligence need to be addressed immediately. How much does the development of artificial intelligence threaten human employment, privacy, equality, and well-being? What controls should we put on killer robots capable of autonomous war? We need an ethical framework to systematically answer all these questions.

Similarly, an ethical framework should provide answers to numerous questions about treatment of animals. Our species is not threatened by other species, but the spread and domination of humans around the world puts at risk the existence of numerous others, including chimpanzees, bonobos, and orangutans. What justifies concluding that such extinctions would be morally wrong, and what can be done to stop them?

Numerous other ethical questions concern the treatment of animals. Are people morally justified in using animals for food, clothing, medical experiments, and other purposes? What is the nature of animal suffering that needs to be taken into account? Are there ethical issues connected with both bots and beasts, such as the construction of hybrids that integrate robots with animals?

To answer these questions, I present a system of morality based on the vital needs of all concerned. Needs provide a way of assessing which bots and beasts have moral standing that make them objects of moral concern. Needs provide ways of addressing existential threats that concern risks of extermination of species such as humans and chimpanzees and also consequential threats concerning less extreme risks to the well-being of relevant individuals.

This framework, combined with relevant facts about current and future developments, generates important conclusions. We do not face the immediate threat of humans being superseded by machine intelligence, but extinction of many intelligent species is a real existential threat. Consideration of needs suggests that killer robots should be banned and that people should stop eating meat. My consideration of the needs of machines and animals will take into account the discussion in chapters 5 and 6 of how humans differ from other forms of intelligence.

Ethics Based on Needs

To answer ethical questions about animals and machines, we need a broad and powerful moral framework. Many people look to religion for answers to ethical questions, but problems arise when we try to apply religious codes to questions about animals and machines. First, our many different religions give different answers to questions concerning animals. Judaism, Christianity, and Islam generally mark humans as inherently superior to other animals, in line with the doctrine of the book of Genesis that God gave humans dominion over animals. In contrast, Hinduism shows much more consideration for animals because of the doctrine of reincarnation: you should treat cows well because they might be your ancestors and you might come back as a cow.

Second, even if religions agreed, we could still ask whether their prescriptions are legitimate. Perhaps the relevant God erred in moral commands, for example, in demanding that Abraham sacrifice his son Isaac. Third, all current religions came into existence long before the existence of intelligent machines, so they have little directly to say with respect to ethical issues concerning robots. Accordingly, we should look for a secular approach to ethics that applies to questions about machines and animals.

The two most prominent approaches to ethics in contemporary philosophy are both potentially relevant to answering serious moral questions. The deontological approach that originated with Immanuel Kant makes ethics a matter of absolute rights and duties. It legitimates asking questions about the rights of animals, the rights of machines, and their impact on the rights and duties of humans. In contrast, the utilitarian approach that originated with Jeremy Bentham ignores rights and duties while considering practices that produce the greatest good for the greatest number of people, where good is measured by pleasure and the avoidance of pain. This approach becomes relevant to animal ethics if we also take into account the pleasure and pain of animals.

Both approaches have well-known philosophical problems. For the deontological approach, where do rights and duties come from? Some people think that rights are God-given, but that claim has the same problem as religious ethics already discussed. Kant tried to establish rights and duties by reasoning alone, but abstract thought helps little to figure out what obligations we have to entities that range from chimpanzees to robots. A standard objection to the utilitarian approach to ethics is that people care about much more than pleasure and pain, so that we should broaden the consequences taken into account in ethical deliberation to include intellectual and social accomplishments.

I prefer an approach to ethics that avoids these pitfalls while incorporating some of the key insights of the deontological and utilitarian approaches. We can figure out what rights people have by considering the vital needs that they must satisfy to lead full human lives. People have biological needs such as air, food, water, and shelter, without which they suffer grievous harm. Psychological research has found that people also have mental needs that include autonomy (freedom from control by others), relatedness (emotional connections with others), and competence (achievement of personal goals such as work and play). Needs differ from petty wants, which can vary from person to person and culture to culture.

In this account, rights derive from needs that are universally important to all human beings. The needs-based approach to ethics takes into account the consequences of different actions for people's happiness but does not narrow happiness down to simple pleasure and pain, because people have many more important goals, such as relationships with other people and achievements in work, play, and other activities. We can answer moral questions in accord with the following Ethical Procedure:

1. List the alternative actions that are worth considering in a particular situation. The ethical deliberation will assess these actions and choose based on moral considerations, not just on personal preferences. For example, government officials can consider whether or not to make military robots more intelligent and autonomous.
2. Identify all the people affected by these actions, including future generations, as well as people alive today. For killer robots, consider people who might be saved, as well as ones who would be killed.
3. For each action, assess the extent to which it helps to promote or impede the satisfaction of human vital needs. For killer robots, the consequences to be considered include the survival and other needs of all the people potentially affected by intelligent weapons.
4. Choose actions that best satisfy human needs.

The major problem with this procedure is that it considers only the needs of humans and ignores those of bots and beasts. So we need to consider the extent to which machines and animals should also be taken into account in ethical deliberations.

Moral Concern

Nothing is morally wrong with chopping up your kitchen table to use for firewood. If the table is yours, you can do with it as you wish. In contrast, universal ethical principles insist that it would be wrong to chop up a person for fuel, food, or fun. Philosophers note this difference by saying that people, but not tables, have moral standing and are objects of moral concern.

Between people and tables are many entities for whom it is more difficult to figure out if they are objects of moral concern. In approximate order of animal complexity, what about bacteria, insects, fish, amphibians, birds, mammals, primates, and apes? Some machines, such as staplers, can openers, and thermostats, seem close to tables in being far from moral concern, but the question becomes more complicated when we consider increasingly intelligent machines such as self-driving cars and future computers with much broader abilities.

Determination of moral standing should be based not on raw intuition but on principles that discriminate between objects of moral concern and other objects. Even with people, the standards have changed over time, as early theories of human rights were largely concerned with the status of propertied white men. Hundreds of years of debate and political action have extended rights and moral concern to all men and women, members of all races, and people independent of sexual orientation and (at least in Canada) gender identity. All such people meet the same standards of moral concern, such as intelligence, self-consciousness, and the capacity to feel pleasure, pain, and emotions.

Basing ethics on needs suggests a different standard: determine whether a class of entities are objects of moral concern by considering their range of needs. Not all apparent needs are relevant, such as when your pencil needs sharpening or your car needs gas. These are not needs of the pencil or the car but rather what people require to make these devices function in accord with their desires. Even in the case of humans, we should distinguish needs from mere wants such as "needing" a BMW to impress your friends.

The needs relevant to ethics are vital ones that are crucial for life and whose nonsatisfaction leads to great harm. You should care about other people because they have the same vital biological and psychological needs that you have. Unless you are among the 1 percent of humans who are psychopaths, you have the capacity to empathize with how humans suffer if their needs are not met. Let us apply the needs-based standard of moral concern to beasts and bots.

Should People Care about Animals?

What are the needs of animals? Consider the pets that are most popular—dogs and cats. These animals have the same biological needs as humans: food, water, air to breathe, shelter from extreme weather conditions, and health care when they are seriously sick. We can expand this set of biological needs to include bodily integrity. A dog or cat that is provided with food, water, housing, and visits to the vet would still be miserable if it were repeatedly beaten by its owner.

Psychological needs are more difficult to attribute to animals. Autonomy is the need to feel that activities are self-chosen to enable people to organize their own behavior following their own interests and values. In chapter 4, I argued that dogs do not have the same sense of self as people. Dogs have interests and values, but little awareness of their own interests and values. Dogs and cats are interested in their food, other animals, and their owners but do not have the cognitive capacity for self-evaluation.

Figure 7.1 charts more than eighty phenomena about the self that have been discussed by psychologists. More than half concern ways in which people represent themselves to themselves and others. One kind of self-representation is evaluation, including appraisals, and criticism, generating emotional products that range from self-admiration to self-loathing. Dogs and cats may legitimately be assigned some simple kinds of self-representation, as when they groom themselves, but we have no evidence that they ever engage in self-evaluation.

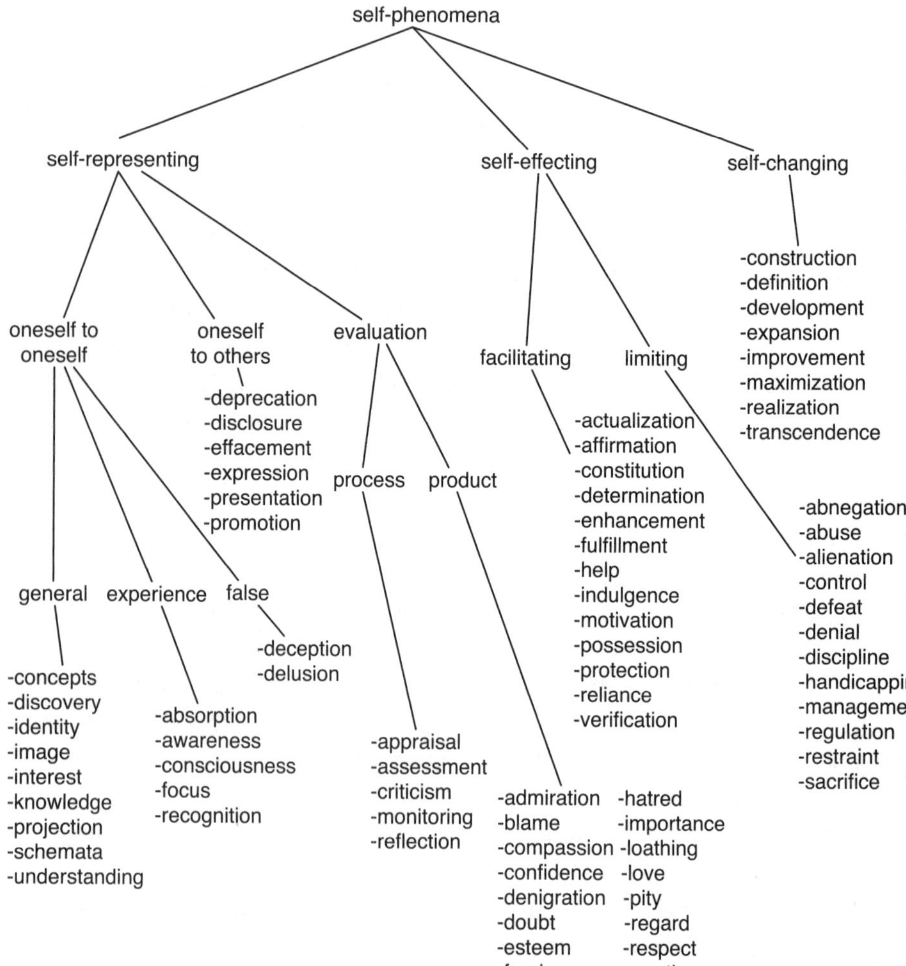

Figure 7.1
Phenomena about the self.

These differences between humans and dogs and the diversity of self-phenomena show that moral standing and moral concern are not binary on/off characteristics but matters of degree that may vary with particular species of animals. We have no need to decide that all animals have exactly the same degree of moral concern, because they have different cognitive capacities with respect to the self. Rather, we can decide for each species, based on its mental capacities, which aspects of moral concern are relevant to it.

Accordingly, we can judge that the human need for autonomy is weaker in animals that do not have the full range of self-phenomena shown in figure 7.1. The phenomena listed under "self-effecting" and "self-changing" show that people have high expectations of their abilities to influence how they accomplish their goals and improve themselves over time. Lacking such expectations, animals such as dogs do not have the same need for autonomy as humans. The same argument applies to animals with greater neurocognitive capacity than dogs, such as ravens, dolphins, and chimpanzees.

Similar differences operate with respect to competence, the need to engage in challenges to display mastery of the world in ways that make people feel effective in their activities. Without self-evaluation, dogs and other animals lack a concern with mastery and effectiveness that goes beyond immediate goals such as finding food.

With respect to the need for relatedness, some species of animals are much more humanlike. Dogs especially have evolved to connect with humans, and even standoffish cats develop bonds with their owners, as the example of jealous Pixie illustrates. Of the animals I have discussed, octopuses are the least social and do not seem to need connections with other octopuses, including their offspring. In contrast, many animals such as wolves and chimpanzees operate in groups. So moral considerations about animals should take into account their need for relatedness while realizing that autonomy and competence may not operate as vital needs for them.

My conclusion is that we should care about animals and consider their needs with respect to biology and relatedness. But they do not deserve the same moral standing and concern as humans, because their lack of the full range of self-phenomena means that they do not have the same psychological needs as humans. There is a hierarchy of mental capacities relevant to establishing moral concern. At the bottom are entities with no life and no sentiments and therefore no needs, including inanimate objects such as rocks and artifacts such as tables. Next are sensing entities such as thermometers, bacteria, and plants, which get information about the world and react to it without having any internal representation of that information.

Animals display more than sensing-reacting, as shown in their experiencing pain, pleasure, and simple emotions. But self-consciousness seems to be much rarer and comes in degrees. The simplest form is body awareness, probably found in many birds and mammals. But this awareness does not add up to the fuller sense of self found in the group of animals that can recognize themselves in mirrors. The highest form of self-consciousness uses all the self-phenomena in figure 7.1, including many forms of self-evaluation and self-change.

In conclusion, animals have vital needs and are therefore objects of moral concern, not just with respect to how they benefit people. But animals' degree of moral standing differs from that of humans because animals have different cognitive capacities and psychological needs. Different animals are capable of different kinds of suffering, distinguished later in this chapter, but their disadvantages disqualify stronger claims such as that chimpanzees are persons.

Should People Care about Machines?

The rock band Queen wrote the song "I'm in Love with My Car," and people sometimes express affection for other machines such as guitars and television sets. But these emotional reactions toward artifacts exist because of what the artifacts do for people, not because the machines are thought to have some special emotional or moral status.

At first glance, the needs-based approach to ethics seems to have a clear stance on questions about the moral standing of intelligent machines. To put it simply, current machines, including the smart ones described in chapter 3, do not have any vital needs because they are not alive. They have no biological needs because they operate without food, water, and air. It is only metaphorical to say that they have a need for electricity, because they do not suffer any harm when the electricity is turned off. Moreover, they lack all the experiences, ranging from pain and emotion to self-consciousness and self-evaluation, that contribute to moral standing, so they have no rights and are irrelevant to moral deliberation. People can do with them whatever they want, including bashing them into pieces and tossing them in a dumpster. Machines lack the characteristics that make bashing and tossing wrong when they are done to people and animals.

Perhaps this could change as machines become smarter. The field of artificial intelligence is less than seventy years old, and I described in chapter 3 how machines have become capable of perception, problem solving, learning, and other features of intelligence. It took four billion years for animal life to go from sensing-reacting bacteria to self-evaluating humans, but machine evolution proceeds much more rapidly.

What would it take for machines to acquire sentience, self-consciousness, and self-evaluation? The last of these is easiest, for it is already feasible to program a computer to evaluate the extent to which it is accomplishing its intended goals, for example, to report its speed of processing and degree of correctness to assigned questions. But such self-evaluation is not relevant to needs unless it is built on self-consciousness that includes an emotional reaction to accomplishing goals.

As I argued in chapters 3 and 6, the path toward machine emotion and self-consciousness is murky, with the best suggestions coming from neurocomputational theories of consciousness. Ideas about neural broadcasting of information, competition among semantic pointers, and information integration all suggest ways in which machines might become conscious by implementing the mechanisms thought to be responsible for human consciousness. The information integration theory of consciousness suggests that machines are already conscious because even smartphones integrate lots of information.

The semantic pointer theory of consciousness suggests, however, that consciousness will not come so easily. For a machine to have a variety of sensory experiences, it requires the property of modal retention, with internal representations that carry forward in reduced form the sensory-motor information that the machine gets from its sensors. Second, for consciousness to serve as a focus of attention, there would need to be a bottleneck in which the representations compete for access to consciousness, although it is possible that future computers will have larger working memory capacity than human brains.

Because of problems that are both technological and motivational, I do not know whether future machines will become sentient and self-conscious in this way. The technological problems are that building machines that use semantic pointers in all the ways humans do will not be easy, and we will probably require much improved theories of human consciousness and therefore even more technological advances. The motivational challenge is whether any computer scientists will actually want to bother to make a machine conscious in this way. I see no obvious commercial or military reasons for making conscious computers, so I do not see any company putting in the financial and human resources to work in that direction.

Another possibility is that consciousness will emerge spontaneously through the self-programming of more and more intelligent computers. But the current state of AI, as found in impressive computers such as IBM Watson and deep learning machines, does not point to computers taking over their own programming. AlphaZero learned mostly by itself to play Go and chess, but it did not write its own learning algorithms. Clever ideas about self-programming have been around for decades, such as John Holland's genetic algorithms based on biological evolution. But genetic algorithms have not produced new kinds of programming, as opposed to solving problems assigned to them by humans.

Although computers will never be alive in the sense of operating with biological mechanisms such as digestion, metabolism, respiration, excretion, and reproduction, they may eventually be conscious. But no sure path leads to that achievement, and today's machines are still far from experiencing pain, emotions, and self-consciousness. Therefore they have no needs and hence no moral standing. So we are not obliged to consider machines in our ethical deliberations except as they affect humans and other animals that do have needs.

Do Machines and Animals Have Moral Obligations?

We have no obligations to machines, so it would be surprising if they had any obligations to us. Even if a computer were programmed to simulate Kantian reasoning about duties or to calculate consequences concerning pleasure, pain, or needs, the machine would not be morally obliged to act in accord with its computations. When people feel they ought to do something, the emotion is a sense of obligation akin to senses of shame and guilt that arise from self-evaluation with respect to moral codes and social relations. So until machines acquire emotions and self-consciousness, we cannot hold them personally responsible for their actions or expect them to have obligations toward us.

The situation is more complicated with animals, as they have needs that make us obligated toward them. Animal researchers have claimed that animals have various characteristics that mark them as moral agents, such as caring, empathy, social learning, and the conscious awareness of appropriateness. Darwin argued that the moral sense of humans differs only in degree from the social instincts of animals to take pleasure in the company of their fellows, feel sympathy for them, and do things for them.

On the other hand, Darwin also claimed that the foundation of morality is the Golden Rule: treat others as you want them to treat you. We find this rule in many world religions, including Christianity, Islam, Buddhism, and Hinduism. It asks you to imagine how you would like other people to behave toward you, so that you will behave analogously toward them. The cognitive complexity of the Golden Rule is clear if you spell it out logically as "For any persons *x* and *y* and any action *z*, *x* ought to do *z* to *y* if and only if *x* wants *y* to do *z* to *x*."

The conclusions about human advantages in chapters 5 and 6 show that no animal can be expected to follow the Golden Rule. First, it involves a high degree of recursion, with looping between what you ought to do to others and what you want others to do to you. No animal has ever been found to be capable of representations with so much recursion. Second, the concept *ought* requires a high degree of abstraction from actual actions toward actions that have the hidden property of being morally obligatory. Human language enables such abstraction through combining concepts, for example, in religion and science, but animals have no such capability. Third, applying the Golden Rule requires a form of analogy in which thinkers imagine their situations as emotionally analogous to those of others and infer how they should act. But even chimpanzees do not seem to be capable of such analogical inferences.

Computers are increasingly capable of these kinds of recursion, abstraction, and analogy, but they are so far incapable of the emotional component of the Golden Rule. Understanding the wants of others requires appreciating the emotional state of desire, which has a physiological aspect that goes beyond the appraisal of goals. Similarly, the concepts of *ought* and *obligation* are familiar to people through feelings that correlate with activity in emotional brain areas such as the ventromedial prefrontal cortex. Whereas animals lack the cognitive capacity to follow the Golden Rule, machines lack the emotional capacity. So neither animals nor machines have moral obligations to humans or anything else based on the Golden Rule.

Existential Threats

From the perspective of needs, we should consider two kinds of threats in assessing the morality of bots and beasts. The most extreme are existential threats that endanger the survival of objects of moral standing. Commentators are increasingly suggesting that machine intelligence poses an existential threat to humans, but I think that such fears are exaggerated. In contrast, the existential threat toward animals is already real, as increasing numbers of species face extinction.

The Robot Apocalypse

Is machine intelligence an existential threat to humans? The issue has been raised in many works of fiction such as the Terminator, Matrix, and Ultron movies. More seriously, it has also been raised by recent commentators such as Elon Musk and Stephen Hawking, who project from recent successes of AI that it is only a matter of time before human intelligence is surpassed by machines. Both the fictional works and personal warnings have been short of information on how human intelligence works and on how close machines are to surpassing it.

At the beginning of chapter 5, I identified numerous advantages that humans have over animals and remarked that most of them are also advantages that humans have over machines. The major exception is the ability to navigate with maps, which driverless cars already do. Some of the other advantages matter little for the question of machine domination over humans because machines could easily do without them. Machines do not need to build fires and cook food because they get their energy electrically and do not care about warmth. Machines can get by without jokes because they do not need humor to facilitate their interactions with humans or other machines.

However, the many other advantages listed in chapter 5 do indicate important ways in which humans are far superior to machines with respect to a wide range of intelligent activities. Philosophical arguments fail to show that machines could *never* close these gaps, but the technological challenges are much greater than can be met just by developing faster computers and more effective machine learning.

Advantages 1 through 6 concern learning and display major gaps between human intelligence and current machine learning. Impressive machine learning algorithms have been developed by humans, but machines are just beginning to produce new learning algorithms themselves. Computers for decades have been able to do much more than simply what they are told to do, because algorithms can exceed human expectations. For example, AlphaZero learned to play Go and chess better than its designers. But so far, machines have worked with the problems that people give them, without the ability to devise new problems to solve.

Human problem solving is not a purely cognitive matter of figuring out how to accomplish goals but also an emotional matter of figuring out what goals are worth accomplishing. Humans are driven by emotions such as curiosity, boredom, greed, envy, and benevolence, all of which provide direction concerning what goals are worth pursuing. Creative humans are capable of generating new goals that others have ignored. They are also capable of inventing new methods such as scientific experimentation, mathematical reasoning, genetic engineering, cubism, and opera. Machine learning from data does not even begin to address questions about the development of new goals and methods. Without such capability, machines would constantly be outwitted by more creative humans if a serious conflict developed between them.

Today's machine learning techniques are excellent at discerning patterns in data and even engage in some kinds of abstraction described in chapter 3. But humans are also capable of abstracting beyond the information in the data to consider hypothetical causes that explain the data. Like animals, computers are excessively empiricist—tied to the observable information that is given to them. In contrast, humans can consider hypothetical causes such as gravity, atoms, genes, and mental representations that could never simply be generalizations from data because they are hidden, nonobservable causes.

Humans are able to do this because of cognitive capacities not yet had by machines: embodied understanding of causality and ability to generate new concepts by conceptual combinations that transcend what is observed. Without causal reasoning about hypothetical entities, humans would never have been able to produce powerful tools such as lasers and effective weapons such as atomic bombs. Until machines understand causality and can use conceptual combination and analogy to generate new hypothetical causes, their scientific and technological abilities will remain a mere shadow of those that humans display.

Humans also have a great advantage over animals and machines in their ability to teach and teach how to teach. Unless all future machine intelligences are identical in hardware and software, they will not be able to teach by instantaneous communication and will have to figure out how to transfer knowledge across systems. We already have computer tutors that teach people skills such as mathematics and computer programming, but I know of no computer tutors for computers. Teaching is also one of the social advantages I discuss later.

Advantages 8 and 9 concern tools and modification of environments. We already have machines using tools ranging from factory robots with welding guns to self-driving cars with cameras and other sensing devices. Moreover, there are factories where robots build other robots, and some research on robots learning to use tools. But I do not know of any machine that uses tools to build the tools that it uses, or a machine that comes close to the ability of humans for looping effects that turns tools into tools into tools. The 3D printers that can produce their own parts fall short of producing tools they can use. Development of layers of metatools will require dramatic improvement in hardware that allows robots to use tools to build tools, and also require more sophisticated software to allow computers to invent and develop plans for tools. To dominate people, machines would need to be able to build metatools that overcome the weapons that people are already able to generate.

Machines would also need to be able to modify their environments and extend the range of habitats in which they can operate, including finding sources of raw materials to regenerate themselves and provide sources of energy. Such modification and extensions will also require advances in hardware and software that are not simply extrapolations from current practices but will require major breakthroughs for flexible operation in complex and changing parts of the world.

Advantages 11 through 13 concern people's emotional abilities to evaluate and improve themselves in ways that have produced many social, medical, and scientific accomplishments. But all the machines in chapter 3, and all the other artificial intelligence research that I am aware of, do not try to use AI for self-evaluation. As already suggested, I do not claim that such procedures are impossible, and I can think of at least one historical AI program, produced by Douglas Lenat, that evaluated heuristics to change themselves. But self-evaluation is rarely a concern of artificial intelligence today, and it would be hard to enact in a general form that would allow the assessment of many different components of an intelligent system, for example, both language capacity and navigation capacity. Humans engage in self-evaluation as part of our emotional systems where we notice what makes us feel good or bad about ourselves. Lacking emotions, machines need some more contrived system based in hardware and software to try to figure out how well they are doing.

Moreover, robots lack the physical capabilities to improve their own bodies. No one has built a machine that can design robots to work better than currently available robots. In addition to the self-evaluation already described, self-modification requires robots to have the manual dexterity to be able to operate on each other, as well as the design capabilities of figuring out how to make robots better. General self-evaluation and self-modification are instances of the overall absence of looping abilities that distinguish humans from animals. As long as machines also lack these capacities, they are little more of a threat to humans than chimpanzees are.

The remaining human advantages, 14 through 18, are all social, reflecting the human capacity to think about other people, collaborate with them, sometimes even risk our lives for them, and modify societies through different forms of government. Some animals are social, but not nearly as communicative and motivated as humans. With rare exceptions such as self-driving cars, most machine intelligence is individualistic. Even when communication among machines occurs, it lacks the gestural and emotional components of human nonverbal communication.

Machines could theoretically work together to try to dominate groups of humans. But without emotions and consciousness, machines would lack the motivation and confidence to pursue domination. The deep learning researcher Yann LeCun observes that "the desire to take over the world is not correlated with intelligence, it's correlated with testosterone." It is easy to prescribe that robots should never have an analogue of testosterone. When humans have horribly set out to dominate other humans in activities such as slavery, war, and genocide, they have been guided by ugly motivations such as greed and xenophobia. Although it is good that computers lack such motivations, it does not augur well for them in a future war that they lack any intense motivations.

Through evaluation of themselves and their societies, humans have been able to develop new modes of communication and government. Social aspirations have produced dramatic technological advances over the last few hundred years, including newspapers, telegrams, radio, television, the internet, and social media. Similarly, social innovations in government practices have improved how people interact with one another through advances such as representative democracy, universal voting, public education, and universal health care. Machine intelligence may not need these particular innovations, but to surpass people it would have to acquire the ability to come up with new ways for computers to interact with and control each other. Otherwise, the immense capacity of humans for social innovation will let us outperform machines unable to improve how they communicate and coordinate with each other.

The mental and social advantages that humans have over animals show that humans are also still enormously more powerful than foreseeable machines. The technologies that make IBM Watson, AlphaZero, and self-driving cars so impressive do not come close to giving computers the looping powers required for producing metatools, self-evaluation, self-improvement, advanced learning, and innovative social coordination.

Surveys of experts find huge variation in when they believe computers might acquire artificial general intelligence that matches humans. For example, in a 2018 survey, predictions about when human-level AI will be achieved ranged from 2029 (Ray Kurzweil) to 2200 (Rodney Brooks, a leading roboticist). None of the AI experts has a theory of human intelligence, whereas my twelve features and eight mechanisms provide the basis for the detailed comparison summarized in the report cards at the end of chapter 3. These comparisons and the human advantages show that we have little reason to fear the robot apocalypse.

Perhaps the consequences of machine domination over humans are so severe that we should already be worrying and taking actions to stop it. Worry could lead to the development of ethical principles that will stop computers from acquiring the looping abilities that would enable them to catch up to people with respect to metatools, self-modification, and other forms of self-improvement. Here are some handicaps that we might want to enshrine in ethical principles that specify what capabilities should be banned from machines: causal understanding, abstraction of hidden causes, looping effects, emotions, consciousness, testosterone, metatools, teaching, creativity with respect to methods, and language with integrated syntax, semantics, and pragmatics. Institutionalizing such handicaps to prevent human-level machine intelligence is a better strategy for preventing the robot apocalypse than trying to figure out how to program a "big red button" into machines that would enable us to turn them off when we are desperately threatened.

Humans now face existential threats much more serious than the robot apocalypse. Global warming is already having severe effects on human life in the form of extreme weather such as storms, droughts, and floods. In twenty years, the effects will be much greater as sea levels rise, the permafrost melts, and ocean currents are redirected. Politicians in powerful countries refuse to act because they are strongly motivated to avoid government restrictions on profitable businesses. Given the substantial human advantages over computers, climate change poses a much more serious threat to humans than machine intelligence does. Deep learning is starting to contribute to global warming because it can require days of computer time for training by examples. Other social problems more pressing than machine intelligence include nuclear war resulting from global conflicts over trade and power, viral pandemics, bacteria resistant to all antibiotics, and growing inequality within and across nations.

Animal Extinctions

Machine intelligence is not an immediate threat to human extinction, but people are a grave threat to the extinction of many other species. Because of human expansion into the habitats of other animals and other environmental effects such as climate change, hundreds of species are becoming extinct every year. Consider the situation for three of the six animals discussed in chapter 4: bees, dolphins, and chimpanzees.

What is wrong with extinction anyway? Most discussions of the robot apocalypse assume that extinction of humans by machines would be bad, although a few transhumanists look forward to our replacement. Almost all the species of animals and plants that have ever existed have gone extinct and been replaced by an abundance of other ones. That our species will eventually go extinct does not directly harm any individual human, because that human will die eventually; so why worry about our species becoming obsolete when each of us is going to be dead anyway? Similarly, why worry about dolphins becoming extinct when there are still lots of other interesting species around?

My conviction that human extinction is undesirable comes from empathy for future generations. People who are not psychopaths are capable of caring about others whom they have never met, for example, victims of natural disasters on the other side of the world. Similarly, people are capable of caring about future generations including not only their own descendants. We can empathize with people living hundreds of years from now, imagining how bad they would feel if machine domination, pandemics, asteroid collisions, or some other disaster is bringing the human species to an end. It is therefore distressing to think that controllable situations such as the rise of intelligent machines could put humans at risk. As climate change intensifies, future generations will think back to us and ask: How could they have been so stupid? Similar resentment could arise if AI starts to become a threat that current actions might have prevented.

Similarly, humans naturally lament the extinction of other animals, both for the sake of humans and for the sake of the animals themselves. Many losses are associated with wide-scale biological extinctions. First, drops in biodiversity can make the planet less flexible in being able to deal with future threats such as climate change. Second, the elimination of animals forecloses opportunities for gaining knowledge about the world. If there are no more great apes, then humans lose the ability to learn from the nature of their closest biological relatives. Third, many humans care about animals as pets or part of nature and will suffer psychologically when those animals are gone.

Is the extinction of the species bad for the species itself? Unlike humans, even the most advanced beasts do not seem capable of conceiving of their species as a whole or imagining far-off future states where that species is extinct. Early death of an animal is of course bad for the animal, but death is inevitable for individuals. So it is hard to make the case that extinction is inherently bad for animal species when they are incapable of imagining the distress of future generations.

More particularly, consider the six smart animals discussed in chapter 4. Fortunately, three of these species are thriving—octopuses, ravens, and dogs. Populations of octopuses and other cephalopods are increasing, thanks to global warming and human fishing practices. Global warming has enabled octopuses to expand their habitats into colder waters, and human catching of predator fish has also allowed octopuses to increase their numbers. Other sea creatures have not been so fortunate, as many species of desirable fish, such as cod, have been seriously depleted.

Ravens have also been doing well because of their flexible diets, which include scavenged carcasses, small animals, and insects. A few other corvids have been threatened, such as the Hawaiian crow, which is now extinct in the wild because of introduced diseases. Many other species of birds have suffered from human encroachment and predation, such as the passenger pigeon, which became extinct because of hunting and deforestation.

Dogs are also thriving because of the growing human population and strong interest in keeping dogs as pets. Hundreds of millions of dogs live in the world, although many of them are wild. In the United States alone, more than seventy million dogs are kept as pets, so dogs face no existential threat.

The ecological situation is much more dire for bees, dolphins, and chimpanzees. In recent decades, bees have been severely threatened by contagious viruses, noxious pesticides, and climate change. Many other species of insects are also threatened in many parts of the world. This decline poses a great threat to human food supplies because most flowering plants are pollinated by insects. Scientists have even warned of the extinction of 40 percent of the world's insect species over the next few decades. Bees lack the mental capacity to justify saying that their extinction would be bad for them as bees, but it would be bad for human welfare because of their contribution to pollination.

Numbers of dolphins are declining because of climate change as warmer waters lead to lower reproductive rates. The Yangtze river dolphin has become extinct because of aggressive fishing practices. Dolphin species members may not be able to form the thought of regrettable extinction, but humans can sadly form this thought and appreciate these animals for their intelligence and grace. Humans would lose out cognitively and emotionally if dolphins became extinct. Fortunately, millions of dolphins are still alive, so they will survive dangers such as being caught in tuna nets for a while.

Much more seriously, all the great apes are in danger of extinction as a result of hunting and invasion of habitat for food production and mining exploration. Chimpanzees are considered an endangered species, with only around two hundred thousand left of the million or so that were alive a century ago. Even more alarmingly, only around ten thousand bonobos may be left, and around a hundred thousand orangutans. The eastern gorilla is down to around five thousand members. Elimination of any of these species robs humans of the opportunity to learn more about our evolutionary history.

Because so many species are threatened by human activities, there is a talk of a sixth extinction, comparable to previous great extinctions such as the one that killed off the dinosaurs at the end of the Cretaceous period. We ought to feel bad about human existential threats to other species for both intellectual and emotional reasons. Such loss of biodiversity impedes the growth of knowledge and the ability of people to appreciate nature.

Consequential Threats: Animals

Even without driving species to extinction, humans treat animals badly in many ways. Cows, pigs, and chickens are raised in crowded factory farms before being killed to serve as food for humans. Most people treat their pet dogs and cats well, but some owners are abusive and end up abandoning their pets to the streets or shelters, where many animals are put down. We can use what has been learned about the minds and needs of animals to address two important ethical questions: whether people should eat animals, and whether people should keep animals as pets.

Should People Eat Animals?

My son Dan has been a vegan since he was sixteen after precociously becoming vegetarian at the age of two. Other early vegetarians that I know adopted the practice of not eating meat around the age of four when they realized that the food chicken was the same as the bird chicken. Dan at two had a visceral revulsion toward meat not shared by his parents, and attempts to hide bits of chicken in his mashed potatoes utterly failed. His later veganism began when he realized that the yogurt I put in his lunches contained gelatin made from horses. In contrast, my younger son Adam likes nothing more than heaps of meat at Brazilian restaurants.

There are four reasons why people should be vegan, avoiding eating meat and dairy products altogether. First, not eating animals avoids the suffering that they experience as a result of being raised for food and then slaughtered. Second, with respect to the two major killers of humans, heart disease and cancer, a vegan diet is healthier than one containing meat. Third, as the human population rapidly approaches eight billion, we have a serious problem of how to feed everyone. Growing large quantities of grains such as corn just to feed animals that are then fed to humans is much less efficient than simply growing plants that people eat. Fourth, raising animals for food is a major contributor to climate change, as the digestion and excrement of cows, pigs, and other animals generate large amounts of methane, which is an even bigger contributor to global warming than carbon dioxide. How strong are these reasons?

The strongest reason for veganism is the following argument: Raising animals for human food causes suffering. Animal suffering is wrong. Therefore raising animals for food is wrong. The first step in evaluating this argument is to figure out what suffering is. Dictionary definitions say that to suffer is to experience pain, distress, or hardship, but a more informative characterization of suffering comes from the method I used for intelligence, causality, and jealousy: standard examples plus typical features plus explanations.

Standard examples of suffering familiar to most people include pain, fear, frustration, exhaustion, sickness, and loneliness. Typical features of suffering are that it occurs when people are consciously aware of an intense and unpleasant experience over a short or long duration. The concept of suffering explains why people behave in particular ways such as moaning, grimacing, or complaining about their situations.

This characterization suffices to show that many animals are capable of suffering. I argued in chapter 4 that birds and mammals are capable of experiencing pain, emotions, and social deprivation. In chapter 5, I contended that fish probably feel pain too. Cows, pigs, sheep, and chickens all have sufficiently powerful brains to be able to suffer, and the Attribution Procedure justifies the conclusion that they often do.

But is it morally wrong if they suffer? Some human beings (narcissists and psychopaths) do not care about the suffering of other human beings, but such people are rare. The 99 percent of people who are not psychopaths have a natural capacity to empathize with other people by imagining their suffering. Over the past few centuries, education and social change have enabled people to expand their circles of empathy beyond their own families, friends, and neighbors to embrace all humanity. Once people understand that animals also have capacities for pain and negative emotions, then empathy should extend to them as well. So animal suffering caused by the actions of people is morally wrong.

Animals might be raised for human food with minimum suffering if they were raised in excellent conditions with ample space, food, and social interactions, followed by a quick and painless death. Today, however, almost all livestock are raised for food in factory farms with little space, miserable living conditions, and deaths that are often clumsy and brutal.

Being a nonvegan vegetarian is more ethical than eating meat, because eggs, milk, and cheese can be gotten from chickens, cows, and sheep without killing them. But the living conditions of egg-laying chickens and dairy cows are scarcely better than those of animals raised for meat, and all these animals eventually encounter slaughtering that is supposed to be humane but is often crudely bungled. So from the perspective of animal suffering, vegetarians are only somewhat better than meat eaters.

Although animals raised for food in factory farms do suffer, their mental limitations probably mean that they do not suffer as much as humans would under such conditions. Humans are capable of a broader range of negative emotions because they have richer representations of selves, more capacity to think of the past and future, a broader understanding of their relations to other people, and the looping capacity for emotions about emotions. These capabilities mean that humans can suffer more than animals because of greater capacity for reflection, memory, imagination, and social emotions; but they do not undercut the conclusion that animals being raised for food in factory farms are suffering.

The second reason for going vegan is more prudential than moral, telling us that we would be healthier if we stopped eating animals. The poster child for this motive is former US president Bill Clinton, who became a near vegan because of serious heart disease. Studies suggest that people who do not eat meat live longer than those who do, although most of the studies involve groups such as the Seventh-day Adventists, whose practitioners may be different from the general population in other dimensions, such as avoiding alcohol.

The responses to the health reason are warnings that vegans may have difficulty getting all the nutrients needed for a healthy diet. Most people who have some knowledge of nutrition can figure out how to get enough protein on a vegan diet, but the issue becomes harder for older people, who need more protein to lessen their chances of becoming frail as their body ages. Getting more protein becomes tricky with age, because a slowing metabolism requires people to cut back on calories, and it is difficult to get a high protein-to-calorie ratio without eating meat and dairy. Other concerns are for nutrients such as iron and vitamin B12, which most people get from meat and fish, but supplements are available. Another issue concerns the omega-3 fatty acid DHA, which is important for maintaining a healthy brain and heart and is most easily obtained by eating fish.

Returning to more ethical questions, the third reason for being vegan concerns the obligation to provide food for the whole world. It takes far more land and resources to raise plants for animals to feed people than simply to raise plants to feed people. In China, for example, rapidly improved prosperity over the past few decades has led to greater consumption of meat and therefore to greater production of plants such as soybeans that are wastefully fed to animals rather than to people. As the human population rises toward the ten billion mark, it will become increasingly difficult to meet the biological needs for feeding all people. Cutting back on meat consumption and production would be a useful way to reduce the chances of widespread starvation.

The fourth reason for being vegan concerns climate change. The prime cause of increased greenhouse gases that are causing global warming is the burning of fossil fuels such as coal, oil, and gas. But animals, especially cows, are also major contributors because their digestion and excretion produce large amounts of methane. My discussion of the robot apocalypse marked climate change as the gravest existential threat to humanity, and dealing with it requires huge changes in energy consumption that could be complemented by changes in eating behavior.

I find compelling the four reasons for being vegan based on suffering, health, feeding the masses, and climate change. Perhaps my son Dan has been right all along, but I have not managed to transition completely to a vegan diet. I continue to eat fish once or twice a week because of evidence that it lowers heart attack rates. I also worry that quitting dairy would make it hard to support the protein needs of my aging body and that becoming vegan would cause social problems in sharing meals with friends who have a more traditional diet.

The arguments from suffering and climate change also count against the use of animals for clothing in items such as shoes, belts, and fur coats. Animal experimentation is more complicated if small amounts of suffering by small numbers of animals can be compensated for by alleviation of much greater amounts of suffering found in human diseases. I have already suggested that human capacities to think about ourselves in the future make us capable of more suffering than animals. But that compensation can only operate if the amount of suffering of animals in experiments is minimal compared to the gains that humans can receive.

Are Pets People?

The slogan "pets are people too" is popular enough to be featured on T-shirts and bumper stickers, and there are books with titles like *Dogs Are*

People Too. The slogan is fine as a way of reminding people that animals have moral standing and deserve to be treated well. But it is problematic both as a matter of fact and as a matter of morality.

My list of eighteen human advantages showed ways in which humans are dramatically different from dogs, cats, and other pets. People's brains are large enough to support looping effects with respect to understanding of the self, social relationships, the future, and morality. Recognizing that animals do have some capacities for emotions, consciousness, and simple language is not enough to elevate animals to people status. Recursive minds have moral properties such as rights and duties that simpler minds lack. Morality depends on empathic appreciation of the needs of others and so is beyond the capacity of even the smartest animals such as chimpanzees and dolphins, let alone cats and dogs.

As a matter of morality, the slogan that pets are people is equally problematic. Slavery used to be widespread in human societies and is still practiced in a few countries, such as Mauritania. But the nineteenth century brought moral progress in the general recognition that one person claiming to own another is deeply wrong. People should not be bought and sold, because every human has a need-based right to autonomy, and slavery also interferes with needs for relatedness (when families are separated) and competence (when people are forced into unsatisfying work). If pets are people, then pet owners are engaging in slavery through practices such as buying, selling, and controlling their pets' environments. Fortunately, the human advantages show that pets are not people, so owners of cats, dogs, and other pets can continue to enjoy their company with a clear conscience.

Some animal activists have proposed other arguments for abolishing pet ownership. The practice leads to the inhumane raising of large numbers of pets in awful facilities such as puppy mills and to the abandonment of many pets to the wild or to shelters where they are often euthanized. Some pet owners do not treat their pets well and in fact increase their suffering. Producing food for carnivorous animals such as cats requires meat production that has all the ethical problems that I discussed with respect to people being vegan.

The countervailing argument in favor of pet ownership is that both owners and pets benefit from a symbiotic relationship. People receive company, affection, and amusement from their pets, who are often treated like family members. When all the needs of pets are satisfied, including food, water, shelter, play, and company, the pets are thriving rather than suffering.

For animals in general, I suggest applying the following test. Would they be better off if they had never been born? One philosopher claimed that people generally fail this test because human lives are miserable, but his claim is contradicted by the fact that most people in prosperous countries are happy. Some animals raised for food do fail the test, for example, calves raised for veal in tiny boxes, and chickens crowded together without enough space to flap their wings. But most pets, like most people, lead lives that are at least pretty good, so their being born was not bad. Policies such as eliminating factory farms and puppy mills would decrease the proportion of animals that would be better off if they had never been born. But it is good that pets are not people, so that people can legitimately own them and take care of them.

Consequential Threats: Machines

Machines lack vital needs, emotions, and suffering, so we should not worry about threats to them, but we should worry about threats from them. Because of the human advantages over current and foreseeable machine intelligence, we can relax about the existential threat of the robot apocalypse. However, machines pose consequential threats to human welfare that are already serious, concerning unemployment, social control, biases leading to social inequality, and killer robots. All these developments are putting humans at risk with respect to satisfaction of their vital needs. I offer a description of these problems, plus suggestions for how society can curtail them.

Automation and Unemployment

From 2015 to 2018, life expectancy in the United States dropped each year in contrast to the previous century of increases. The major cause for this decline was increase in the "diseases of despair": drug overdoses, suicide, and cirrhosis of the liver. Increases in drug use, alcohol, and suicidal despair have largely resulted from losses of jobs in declining occupations such as Appalachian coal mining and Midwestern manufacturing. Mental health is not a purely individual matter but is heavily influenced by changes in society such as unemployment.

Machine intelligence is already having an effect on people's jobs. Manufacturing decline is often blamed on shifting of jobs to low-income countries such as Mexico and China, but it also results from increased automation, with robots replacing humans in factories. Most robots used in factories repetitively do simple tasks and are not particularly intelligent. But advances in machine intelligence will accelerate the process of replacing humans in more challenging work. For example, if self-driving cars and trucks become fully effective, humans will lose millions of driving jobs, including taxis, Uber, Lyft, and truck driving. Other industries threatened by intelligent automation include call centers, retail, and food service. It is possible that losses in such jobs will be compensated for by increases in jobs in other areas, but these may require skills beyond the reach of the workers displaced.

Loss of jobs produces serious harm because of effects on both biological and psychological needs. Without strong social support such as enduring employment insurance, lack of money can lead to inability to satisfy vital physical needs such as food and shelter. In the worst cases, the end of employment can lead to malnutrition and homelessness. Unemployment from intelligent automation can also have severe psychological effects, because people often depend on jobs for satisfaction of their needs for relatedness, autonomy, and competence. Work helps people to satisfy their need for accomplishments and achievements, which cannot be fulfilled by sitting at home and watching television. Moreover, working with friends is one of the ways that people can satisfy needs for social relations. Other sources of relatedness such as families can also be negatively affected by unemployment when despair leads to marital breakdowns and suicide. Unemployment also has a negative impact on satisfaction of the need for autonomy, because people without reliable incomes lose their ability to control their own lives.

Because of these impacts on need satisfaction, we should view social changes resulting from intelligent automation as ethically undesirable. What can we do about this trend? All major companies are currently pursuing artificial intelligence initiatives, and no ways exist to regulate the expansion of intelligent automation within a capitalist economy.

Nevertheless, social policies can help to lessen the impact of employment changes brought about by machine intelligence. Universal health care, including mental health support, can ensure that loss of employment does not eliminate access to medical treatment. Extended unemployment insurance can give the unemployed more time to figure out how to adapt to a rapidly changing economy. My book *Natural Philosophy* gives a needs-based ethical argument for the establishment of a basic income, a payment that guarantees that all members of society can sustain their vital needs.

There are also worthwhile activities that could be publicly funded to provide more employment, for example, ones to improve education and to combat global warming. Major changes in social policies are required to adapt to changes in employment that result from the spread of machine intelligence.

Privacy and Social Control

I have two Amazon Echoes operating in my home that listen for my voice all the time. Amazon assures people that the Echo is only listening for the wake words such as "Alexa," and only when it hears that word does it send the ensuing query to Amazon for interpretation and action. Still, it is unnerving that Amazon could be listening in on everything that I do, from conversations to intimacy. My iPhone and iPad are also waiting for me say, "Hi, Siri," so that it can respond. Such listening makes possible a degree of surveillance that is unprecedented, as companies and governments can monitor communications automatically. Even without speech recognition systems in their homes, people are divulging huge amounts of personal information through their interactions with Facebook, Google Search, Google Maps, and other web programs. For example, if your phone uses apps that track your location, companies can easily figure out how often you go to church or the liquor store.

Companies are already offering shoplifting surveillance driven by machine learning that enables predictions about which people on video cameras might be likely to steal. In China, jaywalkers recognized by machine vision are immediately reprimanded. Companies and governments are increasingly able to invade people's privacy and control their future actions, leading to losses in the satisfaction of the need for autonomy.

Intelligent technologies such as speech recognition, face recognition, natural language understanding, and machine learning have the potential to reduce people's autonomy drastically because organizations can predict what people are likely to do, and subtly control their actions. Intrusions on privacy and increases in social control can also influence satisfaction of needs for competence, because people's ability to accomplish projects they deem valuable could be limited by social control that steers them away from their own interests and toward political priorities. Needs for relatedness are threatened by lack of private communications that are needed to build personal relationships and intimacy.

What can we do about these threats to autonomy, competence, and relatedness? The major forces of technological change today are companies such as Google, Facebook, and Amazon, which have near-monopoly control over their respective internet enterprises of search, social interaction, and shopping. We need new regulations to limit the ways such companies can amass and employ data to invade and control people's lives. Similarly, new legislation should ensure that governments are not using techniques such as face recognition, language understanding, and machine learning to intrude on people's lives. Perhaps new intelligent technologies can be used as tools to monitor the use of such technologies in socially undesirable ways by companies and governments, a looping effect that could provide people with some relief from intrusions by intelligent machines.

Bias and Inequality in Machine Learning and Recommender Systems

You might think that machines would be free of the biases and prejudices that stoke human inequality. A computer cannot be racist and sexist like humans because it lacks emotions such as fear, anger, disgust, contempt, and resentment. Unfortunately, some applications of machine intelligence and natural language processing have shown that computers can incorporate and amplify the unjustified discriminations used by people.

Some companies are trying to employ artificial intelligence to improve their hiring processes. A company that receives thousands of applications for a job faces the problem of sorting through a vast number of applicants to figure out who is likely to best serve the company's goals. Techniques such as natural language processing and machine learning sound perfect for wading through countless résumés and using past experience to predict which people will work best.

But consider the experience of Amazon, revealed in a news report in 2018. Amazon had been trying to use machine learning to improve its hiring by identifying patterns of employees that had worked out in the past to predict who would work out in the future. The problem was that Amazon is a technology company that had mostly hired men, so its new machine learning program was exhibiting a strong bias against women. For example, if the word "women" occurred in a résumé because of mentions of a women's college or club, then the program would be less likely to recommend that person because the female did not fit the pattern of previous hires. So past practices of hiring men rather than women reinforce themselves rather than supporting a serious effort to hire the best candidate. Machine learning algorithms serve to make statistical predictions based on past practice, but past practice may have been discriminatory and not in the interests of the company's desire to hire the best people in the future.

Similarly, a study of the meaning of terms in computer texts found biases similar to those exhibited by humans. Word associations suggest unobjectionably that flowers are more pleasant than insects, but also that European names are more pleasant than African names and that male names are more pleasant than female names.

Bias, prejudice, and discrimination are morally wrong because they interfere with the ability of disadvantaged groups of people to satisfy their biological and psychological needs. If natural language processing and machine learning infer from previous examples that Albanians, for example, are unintelligent, then they will be less likely to be hired in ways that promote their abilities to get food and shelter and to satisfy their psychological needs for autonomy and competence. Not being hired for a job for which you are well qualified limits your freedom and your ability to achieve your goals, as well as blocking social connections with potential coworkers. Discrimination causes harm and suffering, and artificial intelligence is unethical when its use hinders human needs satisfaction.

Avoiding such harm depends on recognizing that technologies such as machine learning and natural language processing are not value-free. Techniques such as deep learning can make it difficult to discern why a program is making the recommendations that it does, because its information is embedded in the connections among thousands of neurons. Unlike a simple rule such as "if the candidate is a woman, then don't hire her," bias in neural networks is much harder to detect. But we have ways of examining a program's performance to determine if it is showing bias against prohibited classes of discriminated people such as women and people of different races. Techniques are being developed to determine whether a program gives biased answers. Social vigilance and political action should serve to monitor the use of intelligent technology so that it can help to reduce rather than increase bias and inequality.

Killer Robots

Almost since its beginnings in the 1950s, the field of artificial intelligence has been funded by the United States Department of Defense Advanced

Research Projects Agency. DARPA funding has had beneficial effects for computer science, such as the development of the internet. But the militaries of the United States, China, and other countries retain a strong interest in using AI and robotics to fight wars. The United States is increasingly turning to robots for reconnaissance and explosives, with plans to have robots in every army formation.

Technologists such as Elon Musk are justifiably alarmed at the prospect of such killer robots. First, combat robots and other devices such as autonomous drones may prove more efficient at killing large numbers of people, eliminating needs satisfaction for all those killed. Second, there are serious ethical questions concerning whether a robot should have the capacity to decide whom to kill. Decisions about killing are bad enough when they are made by humans who have a moral sense of right and wrong, but would be much more cold, automatic, and arbitrary when done by a robot. People can empathize with the needs of others and take those needs into account when making moral decisions.

The simplest way to avoid the ethical problems of killer robots is to prohibit them by insisting that machines should never have the autonomy to choose their victims. In 2017, Musk and 116 AI specialists from twenty-six countries called for the United Nations to ban the development of lethal autonomous weapons. I agree that such a ban is justified because of the potential harm to human needs that could be brought about by robots, drones, and other systems capable of deciding on their own to kill people.

The United States and other countries now developing intelligent weapons will resist the ban, but a weaker alternative would be ineffective. Military technology could aim to program killer robots with the ethical effectiveness of humans or even better, just as self-driving cars have the potential of driving better than people because cars lack distraction and fatigue. But robots will fail to approximate human ethical judgments as long as they lack the emotional capacity for empathically feeling the needs of others and sympathetically caring about them. Until then, robots are at best intelligent psychopaths that should not be trusted with the capacity to kill.

Need, Not Greed

My description of these serious consequential threats of machine intelligence should not give the impression that AI is inherently evil. AI is already helping to improve human society in areas such as language translation, medical diagnosis and treatment, and interactions with digital assistants. Other promising future uses include helping to deal with problems of climate change and facilitating efficient transportation through self-driving cars. All technologies have risks as well as benefits, and the combination of ethical deliberation and public policy development can help to ensure that machine intelligence will be a technology that primarily helps people rather than harming them. In chapter 8, I provide a deeper discussion of principles and values for ethical artificial intelligence.

Mahatma Gandhi said, "The world has enough for everyone's needs, but not everyone's greed." Like other ethical decisions, decisions about AI and ones made by intelligent machines should operate in the service of human needs. Too many decisions today are made in the service of greed for wealth and greed for power. Ethical AI can flourish if it puts universal human needs ahead of the intense, selfish desires that constitute personal, corporate, and political greed.

Similarly, ethical decisions about animals should be based on the needs of both humans and animals rather than on greed manifested in exploitative factory farms and puppy mills. The human advantages compared to animals show that their needs are not the same as humans and that pets are not people. But the cognitive and emotional capacities of animals show that they deserve moral standing, unlike current machines, which are cognitively limited and utterly deficient with respect to pain, suffering, empathy, and social emotions such as shame and guilt.

Implementing ethical rules and algorithms in computers does not accomplish making them care. Values in people are not just preferences or particular goals but rather general goals with emotional attachments. Behaving ethically requires caring about people and values such as vital needs, equality, and democracy. So we cannot align computer values with human values because computers do not have any values, although they might be programmed to fake it.

The cognitive gap between humans and machines has been closing in recent years, as evident in the accomplishments described in chapter 3. But AI is still a long way from achieving DeepMind's plan to "solve intelligence," leaving humans ample time to sort out ethical problems such as killer robots and to establish government and corporate policies that serve human needs.

8 The Ethics of Artificial Intelligence

The dramatic theoretical and practical progress of artificial intelligence in the past decade has raised serious concerns about its ethical consequences. In response, more than eighty organizations have proposed sets of principles for ethical artificial intelligence. The proposed principles overlap in their concern with values such as transparency, justice, fairness, human benefits, avoiding harm, responsibility, and privacy. But no substantive discussion of how principles for ethical AI can be analyzed, justified, and reconciled has taken place. Moreover, the values assumed by these principles have received little analysis and assessment. Perhaps issues about principles and values can be evaded by Stuart Russell's proposal that beneficial AI concerns people's preferences rather than their ethical principles and values.

In this chapter, I argue that the use of principles in medical ethics provides a good model for bringing order to the overabundance of AI proposals. In contrast to the hundreds of principles that have been proposed for AI ethics, a common approach to medical ethics gets by with four key principles concerning autonomy, justice, benefits, and avoiding harm. By sampling current AI proposals, I infer that ethical principles for AI fall under these four principles. Then the considerations that provide legitimacy to the four principles of medical ethics carry over to more specific AI principles.

Ethical principles in medicine and AI presuppose values such as transparency and fairness, but what are these values, and how can they be justified? I argue that values are emotional mental representations that are justified by the extent to which they contribute to the satisfaction of human biological and psychological needs. Then values can be objective as well as psychological and emotional, allowing them to provide legitimate guidance to the practices of artificial intelligence research and applications.

The principles-and-values approach to AI ethics has numerous advantages over trying to build beneficial AI out of human preferences. Practically, it is far easier to work with a handful of principles and values than with billions of preferences. Objectively, principles and values can be judged to be legitimate in ways that avoid the subjectivity of preferences.

Medical Ethics as a Model for AI Ethics

Whereas AI ethics is a new field, medical ethics dates back more than 2,500 years to the Hippocratic oath, and the term "medical ethics" originated in 1803. Ethics is a contentious field of philosophy with many competing approaches, such as religious ethics, virtue ethics, utilitarianism, and Kantian ethics, and this contention spills over into arguments about how to approach medical ethics. Nevertheless, an approach based on four straightforward principles has proved highly useful in theory, applications, and teaching.

In a textbook that has gone through seven editions, Tom Beauchamp and James Childress show the relevance to medical decisions of four principles that can concisely be stated as follows:

(Med1) Autonomy: Respect people's freedom.

(Med2) Beneficence: Provide benefits to people.

(Med3) Nonmaleficence: Avoid harm to people.

(Med4) Justice: Distribute benefits, risks, and costs fairly.

What justifies the adoption and application of this particular set of principles? Beauchamp and Childress view these principles as part of a *common morality* that is universal across humans, but they provide no evidence that the principles are accepted in the thousands of different human cultures. Even in American culture, some of these principles have been challenged, for example, by libertarians such as followers of Ayn Rand. They think that the principles of beneficence and justice are signs of a weakling altruism that threatens the freedom of the supreme individual. The principle of autonomy has been ignored by countless authoritarian regimes in the past and present, and the principle of justice is implicitly challenged by the increasing inequality in most countries of the world. These rejections of the principles show that they are by no means common or universal, and also undermine any claims that they should be accepted because they are self-evident or intuitively obvious. Even if the principles were culturally universal, it would still be legitimate to ask whether they are any good. Correlatively, that principles are not culturally universal does not show that they are wrong, since popular cultures are often ignorant of many truths, such as scientific ones.

Nevertheless, we can identify good reasons why the four principles are appropriate for dealing with ethical problems in medicine. First, they have contributed to informed discussion by numerous authors about pressing issues in medical ethics, including informed consent, euthanasia, surrogate decision making, universal health care, and patient confidentiality. Their application is not simple, because conflicts often arise among principles, for example, when beneficence encourages giving people valuable treatments that they do not want. Nevertheless, the principles provide a framework for deliberation that encourages people to think hard about right and wrong in medicine.

Second, when people have proposed additional principles of medical ethics, they can reasonably be subsumed under the four. For example, truth telling and confidentiality have sometimes been proposed as additions, but these can be justified by their contributions to beneficence, nonmaleficence, and autonomy. The United Nations adopted six "Principles of Medical Ethics Relevant to the Role of Medical Personnel in the Protection of Prisoners," but all apply issues of freedom, benefits, harm, or fairness. For example, principle 1 says:

> Health personnel, particularly physicians, charged with the medical care of prisoners and detainees have a duty to provide them with protection of their physical and mental health and treatment of disease of the same quality and standard as is afforded to those who are not imprisoned or detained.

Concern with health and disease displays beneficence and justice, and other considerations such as the prohibition of torture display nonmaleficence and autonomy.

The third reason for appreciating the four principles of medical ethics is that they efficiently and elegantly incorporate insights from the major approaches to ethical theory: religion, consequences, duties, and virtues. Religious ethics claims that right and wrong derive from divine instructions such as the Ten Commandments, but many of these are instances of beneficence (honor thy father and mother) or nonmaleficence (thou shalt not steal). Utilitarianism is the ethical theory that advocates pursuit of the greatest good for the greatest number, where good means pleasure and the avoidance of pain. Calculating such consequences is a daunting empirical and computational task, but the principles of beneficence and nonmaleficence serve as approximations. Duty-based ethics such as Kant's categorical imperative have been used to justify allowing free actions and acting fairly, which fit with the principles of autonomy and justice. Finally, virtues ethics advises acting in line with personal characteristics such as caring, compassion, integrity, and trustworthiness, all of which contribute to beneficence.

The fourth reason for adopting the four principles is that they fit with the view that the objectivity of ethics derives from the reality of human needs discussed in chapter 7. Vital needs are different from wants and interests in that they are facts about human nature that can be identified by empirical studies concerning requirements for living as a human being. Biological needs are obvious, as people die without oxygen, water, food, shelter, and health care. Psychological needs are more subtle but include relatedness to other people, autonomy in pursuing one's goals, and the competence to achieve goals; substantial research establishes their validity. Satisfying these needs in all humans is promoted by recognizing the goals of beneficence and nonmaleficence and the constraints of autonomy and justice. People need to be treated benevolently and fairly to satisfy their needs.

The final reason for recognizing the four key principles as a good approach to medical ethics is that their application goes far beyond medicine. Doing good, avoiding harm, encouraging freedom, and establishing justice are valuable in many other areas of human activity, such as politics, business, and science. The applicability of the four principles to artificial intelligence can be shown by connecting them to myriad principles recently proposed.

The Asilomar Principles

In January 2017, a conference took place in Asilomar, California, with one hundred participants who included prominent AI researchers and leaders from other fields. After much discussion, they came up with a list of twenty-three Asilomar AI Principles that had the endorsement of at least 90 percent of the participants. I focus on this list because it is longer and broader than those proposed by other organizations and has the endorsement of a large number of people who know AI well. All the twenty-three Asilomar principles fall under at least one of the four principles from medical ethics.

Beneficence

Six of the Asilomar principles clearly fall under the general principle that actions should provide benefits to people.

(Asilomar1) **Research Goal**: The goal of AI research should be to create not undirected intelligence but beneficial intelligence.

(Asilomar2) **Research Funding**: Investments in AI should be accompanied by funding for research on ensuring its beneficial use, including thorny questions in computer science, economics, law, ethics, and social studies.

(Asilomar14) **Shared Benefit**: AI technologies should benefit and empower as many people as possible.

(Asilomar15) **Shared Prosperity**: The economic prosperity created by AI should be shared broadly, to benefit all of humanity.

(Asilomar20) **Importance**: Advanced AI could represent a profound change in the history of life on Earth, and should be planned for and managed with commensurate care and resources.

(Asilomar23) **Common Good**: Superintelligence should only be developed in the service of widely shared ethical ideals, and for the benefit of all humanity rather than one state or organization.

These principles insist that AI research, investment, technology, and prosperity should benefit all humans rather than the financial gain or political power of particular individuals or organizations. The emphasis of principles 14, 15, and 23 on shared benefits introduces elements of fairness, so they also fall under the principle of justice. Superintelligence is the possible development of AI systems more capable than humans and could also have serious consequences for human harm, freedom, and justice.

Two other Asilomar principles fall under the beneficence principle, although the benefits are more indirect.

(Asilomar3) **Science-Policy Link**: There should be constructive and healthy exchange between AI researchers and policy makers.

(Asilomar4) **Research Culture**: A culture of cooperation, trust, and transparency should be fostered among researchers and developers of AI.

The main ethical reason for constructive exchanges between AI researchers and policy makers, and for cooperation among AI researchers and developers, is so that they can all work together toward systems that benefit humans. Such communication should also help deal with possible harms as they arise, making these two principles also relevant to nonmaleficence. I discuss later how the value of transparency is primarily relevant to avoiding harm.

Nonmaleficence

Six of the Asilomar principles are plainly concerned with avoiding specific kinds of harms:

(Asilomar5) **Race Avoidance**: Teams developing AI systems should actively cooperate to avoid corner-cutting on safety standards.

(Asilomar6) **Safety**: AI systems should be safe and secure throughout their operational lifetime, and verifiably so where applicable and feasible.

(Asilomar7) **Failure Transparency**: If an AI system causes harm, it should be possible to ascertain why.

(Asilomar18) **AI Arms Race**: An arms race in lethal autonomous weapons should be avoided.

(Asilomar21) **Risks**: Risks posed by AI systems, especially catastrophic or existential risks, must be subject to planning and mitigation efforts commensurate with their expected impact.

(Asilomar22) **Recursive Self-Improvement**: AI systems designed to recursively self-improve or self-replicate in a manner that could lead to rapidly increasing quality or quantity must be subject to strict safety and control measures.

Principles 5, 6, and 22 concern safety, which aims to prevent harm. Principles 7 and 21 aim to lessen harm by creating transparent systems where the causes of harm can be identified for future prevention and reduction. Principle 18 identifies a particular kind of harm from the increasing use of AI to generate systems that can hurt people without human intervention, for example, in killer robots.

I place another principle under nonmaleficence, although it is vague and requires interpretation.

(Asilomar19) **Capability Caution**: There being no consensus, we should avoid strong assumptions regarding upper limits on future AI capabilities.

I surmise that the main reason for avoiding assumptions about upper limits is to foster awareness of possible harms to people resulting from superintelligence. Ignoring the possibility of superhuman AI would block deliberation concerning potentially serious problems such as people being enslaved, dominated, or rendered pointless by future intelligent computers.

Autonomy

Four of the Asilomar principles are concerned with protecting human freedoms.

(Asilomar12) **Personal Privacy**: People should have the right to access, manage, and control the data they generate, given AI systems' power to analyze and use that data.

(Asilomar13) **Liberty and Privacy**: The application of AI to personal data must not unreasonably curtail people's real or perceived liberty.

(Asilomar16) **Human Control**: Humans should choose how and whether to delegate decisions to AI systems, to accomplish human-chosen objectives.

(Asilomar17) **Nonsubversion**: The power conferred by control of highly advanced AI systems should respect and improve, rather than subvert, the social and civic processes on which the health of society depends.

Privacy, mentioned in principles 12 and 13, is an autonomy issue because violations give AI-using organizations knowledge about people that can be used to restrict their freedom, for example, when media corporations use personal data to manipulate people's behavior, or when governments use private information to identify possible subversives. The right to privacy is also supported by the nonmaleficence principle because people can be harmed when their personal information becomes widely available. Principles 16 and 17 concern loss of human freedom that could result from transfer of control to AI systems or to organizations assisted by AI systems.

Justice

I already mentioned some Asilomar principles that touch on fairness through mentions of sharing, but four Asilomar principles more explicitly concern justice.

(Asilomar8) **Judicial Transparency**: Any involvement by an autonomous system in judicial decision making should provide a satisfactory explanation auditable by a competent human authority.

(Asilomar9) **Responsibility**: Designers and builders of advanced AI systems are stakeholders in the moral implications of their use, misuse, and actions, with a responsibility and opportunity to shape those implications.

(Asilomar10) **Value Alignment**: Highly autonomous AI systems should be designed so that their goals and behaviors can be assured to align with human values throughout their operation.

(Asilomar11) **Human Values**: AI systems should be designed and operated so as to be compatible with ideals of human dignity, rights, freedoms, and cultural diversity.

Principles 8 and 9 insist that AI systems should be sufficiently transparent and accountable that they can be used fairly. Principles 10 and 11 prescribe that AI systems must operate in line with human values such as fairness, although these values could also include elements of doing good, avoiding harm, and promoting freedom.

According to an old joke, a camel is a horse designed by a committee. The twenty-three Asilomar principles were designed by a large committee of committees, so it is not surprising that they lack organization and justification. Fortunately, they are naturally organized under the four key principles originally designed for medical ethics, with eight Asilomar principles primarily falling under beneficence, seven under nonmaleficence, four under autonomy, and four under justice.

Implications

I am not suggesting that the twenty-three Asilomar principles should be replaced by the four used in medical ethics. The twenty-three deal explicitly and wisely with live issues in AI today and are useful because they mention important specific issues such as transparency and safety. Rather, the point of connecting the twenty-three principles to the more parsimonious four is to provide order to what would otherwise seem miscellaneous. Similarly, we can organize other proposed AI principles by connecting them to the key four principles, as I show for other lists of principles in a supplement available at paulthagard.com. Other lists introduce specific concerns such as promoting sustainable environments, but these fit under the key four principles, as I show in my discussion of values.

Organizing the large number of AI principles under the key four from medical ethics has other advantages as well. First, it provides heuristic guidance for how to look for additional principles to deal with specific problems. For example, with respect to the pursuit of AI systems that are smarter than people, we can ask how they would affect human benefits, harms, freedom, and justice.

Second, it is much easier to keep in mind four principles than twenty-three or even ten, so the key four have cognitive advantages. Third, the key four provide a concise implementation of an approach to ethics that needs to be critically compared against alternative approaches, including the following:

1. Skepticism: morals are relative to particular individuals and cultures, so there is no point in worrying about the ethics of AI.
2. Insistence on theory: rather than principles, we should employ a specific theory such as ethics based on religion, virtues, utilitarianism, or Kantian duties and rights.
3. Preferences: we can dodge identifying specific values by focusing instead on what people want as shown by their behaviors.

I argue that the principles approach to ethics is superior to the preferences approach. I ignore skepticism because it has not been advocated by any AI researchers.

Principles versus Preferences

Stuart Russell is a distinguished researcher in artificial intelligence and coauthor of the field's standard textbook. His book *Human Compatible: Artificial Intelligence and the Problem of Control* offers a richly informative discussion of how to design machines with high intelligence that can help people with difficult problems while not making them unhappy. Unfortunately, his preference-based proposal on how to build beneficial AI is impractical and contrary to plausible views in psychology and philosophy.

Russell's proposal is summarized in his three principles for beneficial machines:

1. The machine's only objective is to maximize the realization of human preferences.
2. The machine is initially uncertain about what those preferences are.
3. The ultimate source of information about human preferences is human behavior.

Russell wants to steer clear of judgments about morality by focusing on preferences rather than values. He thinks that it should be possible to prove theorems about how to ensure that AI systems benefit humans in the sense of maximizing their preference satisfaction. Then AI can dodge ethical questions about what specific principles and values should guide the technologies to be developed.

The first problem with Russell's preference approach is practical. He estimates that each person makes twenty trillion choices in a lifetime, and more than seven billion people live in the world, yielding an astonishing number of behaviors that one would have to take into account in inferring preferences from behaviors. Even superfast computers with the best machine learning algorithms would have difficulty mining a huge database of human behaviors for meaningful preferences. If such algorithms did enable computers to predict human behavior with some reliability, there is no guarantee that people could understand the basis for these predictions; techniques such as deep learning in neural networks often yield predictions without clarity about their basis. So learning general preferences from human behaviors is unlikely to yield useful generalizations, let alone courses of action that are provably beneficial. In contrast, much of medical ethics can efficiently be summarized by the four key principles, which are far more informative than billions of behaviors.

The second problem with Russell's approach is that it is based on conceptions of psychology and economics that have been obsolete since the 1960s. Some economists still use the preference-based approach to utility that became popular in the 1940s, when the social sciences were dominated by the behaviorist dogma that it is unscientific to talk about mental representations and processes. But by the 1960s, psychologists had realized the poverty of behaviorism for explaining the actions of humans and other animals and developed cognitive approaches. Similarly, experimental economics has shown that rational choice theory fails to explain human actions and instead looks to emotions and other psychological factors.

From the cognitive perspective, choices and preferences result from mental representations such as goals, emotions, and values. Many medical ethicists and generations of students have found that the values embodied in the four key principles can concisely inform judgments about difficult issues. Old-fashioned economics gets it backward: mental processes cause preferences that cause behaviors, so you cannot just derive preferences from behaviors.

Russell's third problem is philosophical: his approach is too descriptive about human behavior and incapable of yielding defensible normative conclusions about the preferences and behaviors that people ought to have. Human behaviors and preferences are full of immorality and irrationality found in common practices such as violence, war, discrimination, and cruelty. Mining behaviors for preferences may yield advice about AI that is capricious or inimical to human welfare rather than genuinely beneficial for humans. In contrast, the four key principles of medical ethics provide plausible guides to making medicine beneficial for humans, and they apply equally well to AI.

The philosophical position closest to Russell's view is *preference utilitarianism*. Traditional utilitarianism says that good actions are ones that promote the greatest good for the greatest number, where good is understood as promoting pleasure and avoiding pain. In contrast, preference utilitarianism replaces considerations of pleasure and pain with observation of preferences, just as twentieth-century economics replaced a psychological concept of utility with a behavioral one derived from preferences. Critics of preference utilitarianism have pointed out problems with identifying preference satisfaction with human welfare. This identification assumes that people are competently self-interested and operating with true beliefs. Both assumptions are empirically false, as evident from widespread practices such as overeating, substance abuse, and support of autocratic leaders.

Alternative views of welfare advocate maximizing happiness or satisfaction of needs rather than preferences or pleasure and pain. Pluralistic consequentialism is easily summarized by the four key principles that incorporate the values of freedom, fairness, benefits, and avoiding harm. Accordingly, here are suitable replacements for Russell's three principles for ethical machines:

1. The machine's objective is to follow principles of autonomy, justice, beneficence, and nonmaleficence.
2. The machine should operate with the values of human freedom, fairness to humans, human benefits, and avoiding harm to humans.
3. The ultimate source of information about human benefits and harms is empirical evidence about vital human needs.

More generally, when ethical AI applies not only to machine judgments but also to ones made by humans, we can require that decisions be based on a small number of principles, such as the key four, that incorporate the appropriate values as goals. Larger numbers of specific principles, such as the Asilomar ones, can be used to address specific AI issues such as killer robots in line with the key four.

In sum, the principles approach to AI ethics is superior to Russell's preference approach for practical, psychological, and theoretical reasons. Practically, it is much easier to compute the application of a small number of principles than to calculate the relevance of billions of preferences. Psychologically, it is natural for people to apply principles rather than stay at the superficial level of behavioral preferences that are caused by underlying values. Theoretically, the preference approach assumes an approach to utilitarianism that is inferior to others that consider happiness or needs rather than mere preferences.

It might seem that Russell's preference approach to ethics is more democratic than the principles approach, since it takes everyone's interests into account rather than dictating them. But the amalgamation of everyone's preferences into a moral decision inevitably leaves some preferences unsatisfied. For example, if I prefer to play with explosives in a street full of children, then I will feel oppressed when many other people's preferences override my entertainment. Ethical judgments unavoidably suppress some people's behavior when it threatens harm to others. The principles approach makes explicit the trade-offs that are implicit and inevitable in a preference-based approach.

One danger with the principles approach is proliferation of principles, as seen in the more than eighty lists already proposed. We can manage proliferation, however, by connecting the many proposals to the four key principles from medical ethics and by conducting a more fundamental investigation of the nature of values.

Values in Principles

Anna Jobin, Marcello Ienca, and Effy Vayena summarize their study of eighty-four sets of guidelines for ethical AI:

> Our results reveal a global convergence emerging around five ethical principles (transparency, justice and fairness, non-maleficence, responsibility and privacy), with substantive divergence in relation to how these principles are interpreted; why they are deemed important; what issue, domain or actors they pertain to; and how they should be implemented.

This summary confuses principles and values. Principles are rules such as the twenty-three Asilomar prescriptions and the four key prescriptions of medical ethics. They are represented by sentences, whereas values are represented by words such as "transparency." So what Jobin, Ienca, and Vayena are describing are values rather than principles. Values can be converted into principles by embedding them in sentences such as the following, which includes transparency:

(Asilomar4) **Research Culture**: A culture of cooperation, trust, and transparency should be fostered among researchers and developers of AI.

In general, a value *V* can be converted into a principle by creating the sentence "*V* should be pursued" or "Pursue *V*."

Jobin, Ienca, and Vayena's emphasis on five values is based on their occurrence in the eighty-four documents the authors analyze with the following frequencies: transparency (73), justice and fairness (68), nonmaleficence (60), responsibility (60), and privacy (47). Other frequently mentioned values are freedom and autonomy (34), trust (28), sustainability (14), dignity (12), and solidarity (6). Frequency of mention does not guarantee ethical objectivity, since it may just reflect biases of AI culture, so we need to answer the following questions: What are values? What makes values objective rather than subjective? What values ought to guide the development of principles of AI ethics? Can an AI system, not just its designers, have values?

What Are Values?

Asilomar principles 10 and 11 concern the importance of aligning AI systems with human values but do not say what values are. Three possible answers are that values are (1) preferences, (2) ideal concepts, or (3) mental representations. Behaviorists, including many economists, would say that transparency, for example, is a value because people choose transparent situations over nontransparent ones. An AI system for medical decision making where people can tell why the system gives its answers is preferred to one where the reasons for decisions are obscure.

The preference view of values has the same problems I raised with Russell's account of beneficial AI. First, it gets the psychology wrong: people do not just have preferences but have mental states such as concepts and emotions that cause their preferences. Second, it gets the ethics wrong: some preferences such as child pornography are wrong because they conflict with legitimate values such as avoiding harm to children. So we need a deeper view of what values are than simply identifying them with preferences.

Many philosophers since Plato have thought of values as ideal concepts independent of human thinking. Then transparency, privacy, and other values for AI ethics would be like Plato's forms: abstract entities independent of time and space. This view has many problems concerning how we could justify inferring the existence of such heavenly entities, how we could determine which abstract values are legitimate, and how such abstractions could operate in AI systems.

People's values can be understood naturalistically as mental representations that combine concepts and emotions. Concepts generally have associated emotional values, and people feel strongly about values such as democracy. For example, transparency is a concept that operates in people's brains with a positive emotional valence. Contrary to the popular view that emotions are subjective reflections of bodily processes, some theories of emotion recognize that emotions also involve appraisals of the goal relevance of a situation. We can evaluate such appraisals as to whether they accurately assess a situation's impact on goals. For example, fear of diving into a shallow pond is reasonable because of the real danger of breaking one's neck, whereas triskaidekaphobia is unreasonable because no one has ever been harmed by the number 13. So values like transparency combine mental concepts with emotional evaluations that can be more objective than preferences.

Definitions of the word "value" with allusions to similar terms like "merit" and "worth" offer little help in saying what values are. But an account of concepts that is more psychologically and neurologically plausible than the traditional definition-based view sees them as combining three aspects: exemplars (standard examples), stereotypes (typical features), and explanations.

Accordingly, we can characterize values by the method used for intelligence in chapter 2 with standard examples, typical features, and explanations. Commonly accepted examples of values include freedom, happiness, life, equality, democracy, health, and work. These values have the typical features of being concepts with positive emotions and the capability to influence judgments and lead to action, with varying degrees of importance. Having such values among people's mental representations explains their judgments and behaviors. With this account of what values are, we can consider what values are legitimate for evaluating AI and for incorporating into AI systems.

What Are the Best Values in General?

One of the justifications I gave for adopting the four key principles of medical ethics is that they promote human needs, both biological and psychological. A similar justification applies to adopting the values of freedom, fairness, benefits, and avoiding harm to them. Each is a concept toward which most people have positive emotions, for good reason. Having freedom enables people to achieve the goals that matter to them rather than being coerced or restricted. Fairness is important for ensuring that goal achievement is not restricted to a privileged few. People's activities are oriented toward getting benefits rather than harms. We can extend this list of values either by showing their connection to the key four or by independently evaluating how they contribute to satisfying human needs.

Values for Developing Ethical AI

Of the eleven values mentioned by Jobin, Ienca, and Vayena, I have already incorporated four within the principles framework: justice and fairness, nonmaleficence, beneficence, and freedom and autonomy. This inclusion leaves transparency, responsibility, privacy, trust, sustainability, dignity, and solidarity. Their legitimacy derives from the four core values.

Transparency in AI includes practices such as making AI systems explainable so that users can understood why they do what they do, and communicating to the general public how such systems work. Such practices are primarily justified by the nonmaleficence value, because people can be harmed if machines are used to hide their purposes. For example, an AI system that purports to provide movie recommendations is not transparent if it is surreptitiously tracking people's values for political purposes. Transparency can also provide human benefits if it enables people to pursue their own goals more effectively, which also enhances their autonomy. Finally, transparency can promote justice if making AI systems understandable and publicly communicating the information makes it easier to share their benefits with all relevant people.

Responsibility is a derivative value because it is exercised by having people apply principles based on appropriate values. AI developers and companies are responsible if and only if they act to produce human benefits, avoid harms, and foster freedom and justice.

Jurists and politicians since Louis Brandeis have defended a right to privacy construed as a right to be left alone without invasive scrutiny by other people and organizations. Privacy contributes to freedom by not allowing others to have information that can be used for coercive purposes. It contributes to benefits because lack of coercion allows people to pursue their own goals, such as having personal relationships, and it contributes to avoiding harm by preventing embarrassment and fear of control by others. Finally, privacy contributes to justice by ensuring that people are treated equally and fairly independent of personal details.

Sustainability of the environment is clearly a matter of benefits and harms. If AI is used to waste energy and contribute to climate change, then it harms humanity, whereas if AI provides new ways of promoting energy efficiency and reducing greenhouse gas emotions, then it can benefit billions of people.

Solidarity in AI principles largely concerns having a social safety net and ensuring that people can engage together in productive work. It is therefore a value that derives from its contributions to human benefits and avoiding the harms caused by loss of freedom.

Dignity concerns being respected and appreciated by others and provides benefits to those who receive it from their peers. Having dignity also avoids harms such as the negative emotions that result from insults, and loss of dignity usually goes with loss of freedom as the result of abuse and coercion.

Trust is an important part of human relationships but as a value is derivative from the benefits that come from people working together productively and getting along. Failures of trust are harmful in that they generate negative emotions such as fear and anxiety.

These assessments show the plausibility of dividing the eleven values into two classes: the four core values that appear in the four key principles in medical ethics, and the seven derivative values that are useful for generating principles that are subsidiary to the key four. Subsidiary principles and values are legitimate and useful for providing guidance in specific areas of interest such as the use of AI programs to threaten privacy.

Figure 8.1 shows the relation among the eleven values, using my method of cognitive-affective maps that has been widely applied to political and ethical issues. I place the four core values inherent in the four key principles at the center of the map with thick ovals. Some of the relations between them and seven other values are shown by solid lines, and incompatibilities with some negative values are shown by dotted lines.

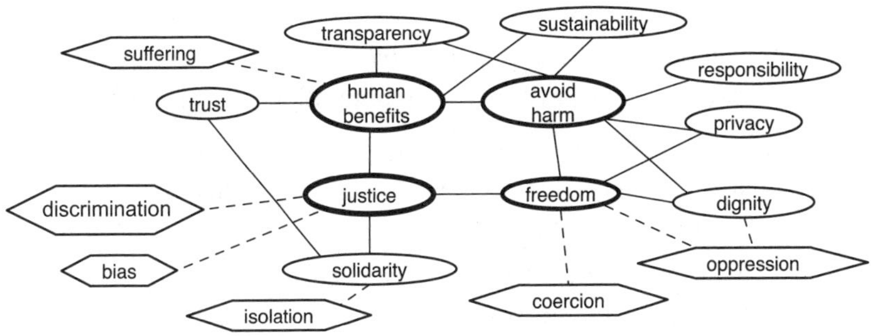

Figure 8.1
Map of values relevant to ethical AI. Positive values are in ovals with thick lines indicating the core; negative values are in hexagons. Solid lines indicate mutual support, and dotted lines indicate incompatibility. Not all values and connections are shown.

Implementing Ethics in AI Systems

Identifying principles and values for ethical AI is useful for two reasons. First, people making decisions about AI can use principles and values to ensure that they are acting ethically. Second, AI systems can be programmed to act ethically if they incorporate principles, values, and other ethical standards, such as cases of good and bad behavior, to serve as analogies. What are the prospects for making AI machines act ethically by themselves?

Computing Principles

For an AI system to make ethical inferences, it needs to have more than just natural language representations of ethical principles such as "Respect people's freedom" (Med1) and "The goal of AI research should be to create not undirected intelligence but beneficial intelligence" (Asilomar1). Since the origins of AI in the 1950s, one of the main methods of representing and processing information has been production rules, also called just productions or rules. Such rules are if-then structures that control inference by processes that include matching the if-part and executing the then-part. For example, the advice to scratch an itch can be converted into the rule "if itch, then scratch," which first matches a situation such as an itchy back and then executes an instruction to scratch it. Production rules have been central to important AI programs such as GPS, ACT-R, and SOAR.

With ingenuity, all ethical principles can be converted into production rules that include the four key principles of medical ethics and the first four Asilomar principles. Here are provisional translations:

(Med1) Autonomy: Respect people's freedom.

If an action affects other people, then set the goal that the action does not restrict their freedom.

(Med2) Beneficence: Provide benefits to people.

If an action affects other people, then set the goal that the action provides benefits to people.

(Med3) Nonmaleficence: Avoid harm to people.

If an action affects other people, then set the goal that the action does not harm them.

(Med4) Justice: Distribute benefits, risks, and costs fairly.

If an action distributes resources among people, then set the goal that the distribution is fair.

(Asilomar1) **Research Goal**: The goal of AI research should be to create not undirected intelligence but beneficial intelligence.

If an activity is AI research, then set the goal that intelligence benefits humans.

(Asilomar2) **Research Funding**: Investments in AI should be accompanied by funding for research on ensuring its beneficial use.

If an investment is in AI, then set the goal of funding research about beneficial use.

(Asilomar3) **Science-Policy Link**: There should be constructive and healthy exchange between AI researchers and policy makers.

If X is an AI researcher and Y is a policy maker, then set the goal that X and Y exchange information.

(Asilomar4) **Research Culture**: A culture of cooperation, trust, and transparency should be fostered among researchers and developers of AI.

If X and Y are researchers or developers of AI, then set the goals that X cooperates, trusts, and is transparent with Y.

Making inferences with production rules is more complicated than just applying modus ponens: If *P* then *Q*; *P*; therefore *Q*. There may be many productions that potentially apply in a situation that would yield contradictory actions. Different production rule systems use different principles for deciding what rules to apply, with possibilities such as applying the more specific rule or applying the rule that has been most successful in previous problem solving. Production rules should be understood as defaults rather than as universal prescriptions.

It therefore is feasible to encode ethical principles into AI systems that make inferences with production rules. For a neural network implementation, rules can be translated into vectors and then into spiking neurons. The two major problems for production rule implementations of ethical principles are coming up with sufficiently precise translations of the rules to allow them to be processed computationally, and dealing with conflicts among rules. Ethical dilemmas are common and involve conflicts among principles. For example, suppose a terrorist has placed a bomb set to explode and kill hundreds of people. Is it legitimate to torture the terrorist to find the bomb's location? Answering this question involves conflicting rules about freedom, benefits, harms, and justice. Standard ways of choosing what rules to apply in production systems provide no guidance about how to balance these concerns. Such balancing may be easier to apply to values.

Computing Values

The account of values as mental representations that combine concepts and emotions is problematic for AI systems because no such systems currently have emotions. If emotions were just cognitive appraisals, then it would be easy for AI systems to have them, and computational models of emotions as appraisals already exist. But emotions involve physiological changes and conscious feelings as well as appraisals. Computers lack anything like the relevant physiological states of humans, such as heart rates, breathing rates, and cortisol and hormone levels that influence emotions. Moreover, no current or planned computers are conscious.

Because values have an emotional component that includes physiology and consciousness, we have no immediate prospect of fully incorporating values into an AI system. But approximations are possible that can allow AI systems to have representations that carry some of the functionality of values.

The simplest approximation is to treat values as unstructured symbols that contribute to the processing capability of more complex representations. For example, my translations of ethical principles such as Med1 and Asilomar1 into production rules employ symbols such as "freedom" and "benefits" that can be taken as stand-ins for the values of freedom and human benefit.

A richer representation of a value such as freedom could be provided using a structure such as the frames of Marvin Minsky, which provide slots that indicate defaults. The frame for freedom might include slots and fillers such as the following:

Power: act, speak, think

Absence of: coercion, restriction

Valence: positive

A richer account of the concept of freedom could incorporate the additional elements that were specified in my characterization of value. Then to the typical features included in a Minsky frame we could add standard examples of freedom, such as Nelson Mandela after his release from a South African prison. In addition, the explanatory use of the concept of freedom could note that it explains the happiness of people free from coercion.

The usual function of such structured concepts is categorization, where objects or situations are classified based on degree of match. For example, a situation could be classified as free depending on how well it matches the slots for power, absence, and valence. Often, however, we need to make an ethical judgment about a more complex situation that matches more than one value.

A clue about how to provide such trade-offs comes from the kind of value network shown in figure 8.1. The value maps are based on a theory of emotional coherence where judgments can be made computationally by treating the values as artificial neurons in a network, where the solid lines indicate excitatory links, and the dotted lines indicate inhibitory links. Then efficient algorithms for parallel constraint satisfaction activate some elements and deactivate others to make a decision. For example, if the artificial neuron for *freedom* is activated, then the neuron for *oppression* is deactivated. Parallel constraint satisfaction can also be performed by methods other than neural networks, such as a greedy algorithm that approximates an optimal solution.

Without emotions, computers lack a full understanding of values and other key ethical ideas such as empathy. But approximations to the cognitive functioning of values can be gained through methods such as symbols in production rules, frame-like representations, and processes for parallel constraint satisfaction.

These methods for computing principles and values are not the only means by which a machine could incorporate ethics. Traditional utilitarianism could be implemented by using a computer to maximize expected utility for all relevant people, using probabilities and utilities in line with economic theories of decision making to calculate the greatest good for the greatest number. This method is difficult to implement given lack of information about probabilities and utilities.

Medical ethics employ an alternative to general principles and values known as "casuistry" or the "method of cases." Instead of applying general principles, ethical reasoning could operate by comparing problematic situations with previous cases whose moral significance is clearer. AI systems could employ this method by using well-known algorithms for analogical inference and case-based reasoning. Analogies, however, have limited use in establishing ethical conclusions about AI because of problems in finding relevant cases and adjudicating among cases that support different interpretations.

Ethical AI

It is commendable that many organizations have been sufficiently concerned with the social consequences of artificial intelligence that they have generated lists of principles to guide research, applications, and public policy. The principles approach to AI ethics has clear advantages over alternatives such as using preferences, analogies, religion, or skeptical relativism.

This chapter advances the principles approach to AI ethics in several ways. First, it organizes numerous and diverse AI principles by using four clear principles that are widely used in medical ethics. The comprehensiveness of these key principles is evident from how they group the twenty-three Asilomar principles. Second, the four principle-related values of beneficence, nonmaleficence, autonomy, and justice provide a framework for analyzing the eleven values that are most prominent in the eighty-four lists of principles analyzed by Jobin, Ienca, and Vayena. Third, principles and values for ethical AI can be incorporated into AI systems by using well-known techniques such as production rules, frames, and parallel constraint satisfaction.

Much work remains to flesh out the four key principles and core values that I have used to organize the ethics of AI. Beneficence and nonmaleficence require an account of benefits and harms that is biologically and psychologically deeper than preferences, pleasure, and pain. We can provide such an account by attending to the nature of human vital needs rather than wants. We can consider the costs and benefits of current and future

AI by considering facts about bodies and minds that enable humans to be human. Moreover, these facts provide insights into how the social contributions of AI can be applied justly to humans irrespective of race, nationality, gender, and other sources of discrimination. Ethical AI can thrive if it puts universal human needs ahead of personal greed in identifying and applying powerful principles and values.

Policy Recommendations

I have systematically compared intelligence in machines, humans, and other animals by assessing their features and mechanisms. The results have implications for the study of intelligence, the treatment of animals, and the development of artificial intelligence. Fifty years from now, if we have managed to survive climate disasters, pandemics, and evil leaders, we should have a much deeper understanding of all kinds of intelligence and have developed new technologies for augmenting, but not replacing, human intelligence.

Investigation of Intelligence Should Be Interdisciplinary

The field of psychology has made many valuable contributions to understanding intelligence, but its narrow focus has had unfortunate consequences. Fixation on IQ and *g* has skewed the concept of intelligence toward linguistic and mathematical skills and away from valuable nonverbal aspects such as emotions and spatial navigation. Mathematical techniques that look for associations among measurable variables have discouraged the discovery of underlying mechanisms that provide deeper explanations. Concentrating on IQ and correlational methods has sometimes enabled arguments intended to undermine racial equality.

The best way to broaden psychological investigation of intelligence is to combine it with ideas and methods from other fields, in line with the interdisciplinary aims of cognitive science. Artificial intelligence provides computational models of problem solving, learning, and the other features of intelligence that serve to identify mechanisms responsible for human thinking. Neuroscience also contributes to explanatory mechanisms, ranging from molecular processes such as neurotransmitters to the interactions of networked brain areas. Ethology adds evidence and theories about intelligence in nonhuman animals. Anthropology ensures that accounts of intelligence are valid across diverse cultures. Because so much of human intelligence is language oriented, the field of linguistics also provides insights. Finally, philosophy has much to contribute to the understanding of intelligence through general reflections on interdisciplinary issues, and especially through examination of the ethical implications of developments concerning machines, animals, and people.

The Needs and Capacities of Animals Should Be Respected but Not Exaggerated

Consideration of the ethical treatment of animals should avoid romantic assumptions that animals are morally equivalent to humans, but also shun killjoy skepticism about their capacities. The Attribution Procedure builds on substantial experimental evidence to provide good reasons why many kinds of animals are capable of consciousness, including degrees of pain, suffering, and emotion. This capability implies that animals have moral standing that should constrain their uses for food and medical experimentation. Human encroachment on natural habitats is threatening the extinction of many species and is especially tragic in the case of the great apes, whose mental capacities are close to those of humans in many respects.

However, the substantial mental advantages of humans show that pets are not people, so that their companionship does not amount to slavery. Well-treated animals whose needs are satisfied can legitimately contribute to human happiness. We must continue to examine the extent to which the limitations of animals with respect to cognition and emotion justify their uses for food and medical information. I would argue, for example, that bees do not suffer from the production of honey.

Artificial Intelligence Should Be Closely Monitored to Ensure That It Satisfies Human Needs

Intelligent machines will increasingly change human lives, but their many limitations with respect to cognition, emotion, and consciousness show that we can postpone worries about whether they deserve moral standing. Similarly, we have no immediate prospects for the development of artificial general intelligence, so we can calm fears about the prospect of a robot apocalypse that subordinates people to computers. Nevertheless, ongoing reflection on the prospects and threats of human-level AI is worthwhile.

It is not too soon to identify ways of handicapping AI to prevent it from usurping humans, for example, by limiting causal understanding, emotions, and creativity.

More immediately, technological developments are already posing serious threats to human well-being. Killer robots are increasingly feasible but should be banned internationally because they violate the ethical principles of avoiding harm to people and maintaining their autonomy. These principles also justify close monitoring of ways in which intelligent computers can make biased inferences that increase inequality. AI can also reduce freedom through violations of privacy that foster social control by governments and corporations. New applications such as driverless cars and automated decision making can thwart the productive employment that contributes to human needs for accomplishment and social connection.

My main overall recommendation is summarized in the slogan "need, not greed." Computers and robots today do not have any needs, so they do not figure in our ethical deliberations. But some animals have psychological as well as biological needs that we should take into account. More strongly, the features and mechanisms of human intelligence generate complex psychological needs for autonomy, competence, and relatedness. Future treatment of animals and computers should be driven by considerations of biological and psychological needs, not by the greed for wealth and power that operates in selfish individuals and uncaring organizations. Compassionate people can work with bots and beasts to promote flourishing for all.

Acknowledgments

My interest in artificial intelligence began long ago with the writings of Marvin Minsky. Much more recently, my interest in animal cognition was stirred by Ashley Keefner's MA and PhD theses on animal minds. I have benefited from conversations with her and many others, including Laurette Larocque, Chris Eliasmith, Yan Wu, Timothy Lillicrap, Iason Daniels, and students in my University of Waterloo course on intelligence in machines, people, and other animals. Laurette provided inspiration, encouragement, part of the title, and cats.

For comments on an earlier draft, I am grateful to Erik T. Mueller, Jim Bowers, Yan Wu, and two anonymous reviewers.

I have used excerpts (for which I hold copyright) from my Aeon.co article "Green-Eyed Pets." I have also used material (for which I hold copyright) from my *Psychology Today* blog *Hot Thought*.

Thanks to Philip Laughlin, Alex Hoopes, and Judith Feldmann for efficient editorial support, to Bill Henry for excellent copyediting, and to Leonard Rosenbaum for thorough indexing.

Notes

See paulthagard.com for live web links, a reference list in APA style, and the supplement to chapter 8.

Chapter 1: Attributing Minds to Machines and Animals

On animals that recognize themselves in mirrors, see H. Prior, A. Schwarz, and O. Güntürkün, "Mirror-Induced Representation in Magpies: Evidence of Self-Recognition," *PLOS Biology* 6, no. 8 (2008): e202.

R. Kurzweil, in *The Singularity Is Near: When Humans Transcend Biology* (New York: Viking, 2005), lauds the singularity. Warnings about the consequences of AI have come from Bill Gates, quoted in H. Pettit, "Killer Computers," *The Sun* (UK), March 21, 2019, https://www.thesun.co.uk/tech/8688058/bill-gates-warning-dangerous-ai-nuclear-weapons; S. Hawking, *Brief Answers to the Big Questions* (New York: Bantam, 2018); Elon Musk, quoted in K. Piper, "Why Elon Musk Fears Artificial Intelligence," *Vox*, November 2, 2018, https://www.vox.com/future-perfect/2018/11/2/18053418/elon-musk-artificial-intelligence-google-deepmind-openai.

The Nature of Human Intelligence, edited by R. J. Sternberg (Cambridge: Cambridge University Press, 2018), contains conflicting definitions of intelligence. Psychological research on concepts is reviewed in G. L. Murphy, *The Big Book of Concepts* (Cambridge, MA: MIT Press, 2002); and L. J. Rips, E. E. Smith, and D. L. Medin, "Concepts and Categories: Memory, Meaning, and Metaphysics," in *Oxford Handbook of Thinking and Reasoning*, edited by K. J. Holyoak and R. G. Morrison (Oxford: Oxford University Press, 2012), 177–209. I call conceptual analysis using exemplars, features, and explanations "three-analysis": see my *Brain-Mind: From Neurons to Consciousness and Creativity* (New York: Oxford University Press, 2019); and *Natural*

Philosophy: From Social Brains to Knowledge, Reality, Morality, and Beauty (New York: Oxford University Press, 2019).

M. Tegmark defines intelligence as the "ability to accomplish complex goals," which packs a lot into "complex," for example, that learning is required; see M. Tegmark, *Life 3.0: Being Human in the Age of Artificial Intelligence* (New York: Alfred A. Knopf, 2017), 39. Psychologists often cite the following definition by Linda Gottfredson: "Intelligence is a very general mental capability that, among other things, involves the ability to reason, plan, solve problems, think abstractly, comprehend complex ideas, learn quickly and learn from experience"; see L. Gottfredson, "Mainstream Science on Intelligence: An Editorial with 52 Signatories, History, and Bibliography," *Intelligence* 24 (1997): 13. The phrase "among other things" makes the definition indeterminate.

For references on machines, see the notes to chapter 3; for references on animals, see the notes to chapter 4.

The "romantic/killjoy" distinction comes from K. Andrews, *How to Study Animal Minds* (Cambridge: Cambridge University Press, 2020); D. C. Dennett, "Intentional Systems in Cognitive Ethology: The 'Panglossian Paradigm' Defended," *Behavioral and Brain Sciences* 6, no. 3 (1983): 343–390; and T. Suddendorf, *The Gap: The Science of What Separates Us from Other Animals* (New York: Basic Books, 2013). S. J. Shettleworth evaluates killjoy explanations in "Clever Animals and Killjoy Explanations in Comparative Psychology," *Trends in Cognitive Sciences* 14, no. 11 (2010): 477–481. Philosophical arguments against the possibility of artificial intelligence include claims that computers will never be able to handle context (H. L. Dreyfus, *What Computers Still Can't Do*, 3rd ed. [Cambridge, MA: MIT Press, 1992]); semantics (J. Searle, "Minds, Brains, and Programs," *Behavioral and Brain Sciences* 3 [1980]: 417–424); significance (John Haugeland, *Having Thought: Essays in the Metaphysics of Mind* [Cambridge, MA: Harvard University Press, 1998]); and creativity (S. D. Kelly, "A Philosopher Argues That an AI Can't Be an Artist: Creativity Is, and Always Will Be, a Human Endeavor," *MIT Technology Review*, March–April 2019, https://www.technologyreview.com/s/612913/a-philosopher-argues-that-an-ai-can-never-be-an-artist). I take these issues as challenges rather than impossibilities; see chapters 3 and 5.

The Attribution Procedure is an application of a theory of explanatory coherence that accounts for the reasoning of Newton, Darwin, Einstein, and many other scientists: see my *Conceptual Revolutions* (Princeton, NJ: Princeton University Press, 1992); *How Scientists Explain Disease* (Princeton,

NJ: Princeton University Press, 1999); *The Cognitive Science of Science: Explanation, Discovery, and Conceptual Change* (Cambridge, MA: MIT Press, 2012). Explanatory coherence can efficiently be computed by neural networks and other algorithms. For a description of inference to the best explanation, see G. Harman, *Thought* (Princeton, NJ: Princeton University Press, 1973); and P. Lipton, *Inference to the Best Explanation*, 2nd ed. (London: Routledge, 2004), which K. Andrews applies to animal cognition in *The Animal Mind: An Introduction to the Philosophy of Animal Cognition* (Abingdon: Routledge, 2015). Explanatory coherence and simplicity show the implausibility of claims that people are living in a computer simulation.

The Morgan quotes come from C. L. Morgan, *An Introduction to Comparative Psychology* (London: Walter Scott Publishing, 1903), 53, 59. On evolutionary simplicity, see F. B. M. de Waal, *Are We Smart Enough to Know How Smart Animals Are?* (New York: Norton, 2016), 43. Mike Dacey analyzes anthropomorphism as a cognitive bias in "Anthropomorphism as Cognitive Bias," *Philosophy of Science* 84, no. 5 (2017): 1152–1164.

Chapter 2: Prodigious People

I use the convention that the word "tree" stands for the thing tree, which is represented in the mind by the concept *tree*. Similarly, the fact that trees are green is represented in text by the sentence "Trees are green" and in minds by the rule *If something is a tree, then it is green*.

Cognitive science approaches to mental mechanisms are reviewed in W. Bechtel, *Mental Mechanisms: Philosophical Perspectives on Cognitive Neuroscience* (New York: Routledge, 2008); M. Boden, *Mind as Machine: A History of Cognitive Science* (Oxford: Oxford University Press, 2006); K. J. Holyoak and R. G. Morrison, eds., *The Oxford Handbook of Thinking and Reasoning* (New York: Oxford University Press, 2012); and myself in *Mind: Introduction to Cognitive Science*, 2nd ed. (Cambridge, MA: MIT Press, 2005), and *Brain-Mind: From Neurons to Consciousness and Creativity* (New York: Oxford University Press, 2019).

Psychological research on intelligence is reviewed in E. Hunt, *Human Intelligence* (Cambridge: Cambridge University Press, 2010); N. J. Mackintosh, *IQ and Human Intelligence*, 2nd ed. (Oxford: Oxford University Press, 2011); R. E. Nisbett, *Intelligence and How We Get It* (New York: Norton, 2009); R. E. Nisbett, J. Aronson, C. Blair, W. Dickens, J. Flynn, D. F. Halpern, and E. Turkheimer,

"Intelligence: New Findings and Theoretical Developments," *American Psychologist* 67, no. 2 (2012): 130–159; R. J. Sternberg, ed., *The Nature of Human Intelligence* (Cambridge: Cambridge University Press, 2018); and R. J. Sternberg and S. B. Kaufman, eds., *The Cambridge Handbook of Intelligence* (Cambridge: Cambridge University Press, 2011). Neural theories of intelligence include R. E. Jung and R. J. Haier, "The Parieto-frontal Integration Theory (P-FIT) of Intelligence: Converging Neuroimaging Evidence," *Behavioral and Brain Sciences* 30, no. 2 (2007): 135–154; and A. O. Savi, M. Marsman, H. L. J. van der Maas, and G. K. J. Maris, "The Wiring of Intelligence," *Perspectives on Psychological Science* 14, no. 6 (2019): 1034–1061. Howard Gardner defends multiple kinds of intelligence in *Multiple Intelligences: New Horizons* (New York: Basic Books, 2006). J. D. Mayer, P. Salovey, and D. R. Caruso review emotional intelligence in "Emotional Intelligence: Theory, Findings, and Implications," *Psychological Inquiry* 15, no. 3 (2004): 197–215.

Neuroscientific research on intelligence connects it to genes (R. J. Haier, *The Neuroscience of Intelligence* [New York: Cambridge University Press, 2016]); epigenetics (J. A. Kaminsky, F. Schlagenhauf, M. Rapp, S. Awasthi, B. Ruggeri, L. Deserno, T. Banaschewski, et al., "Epigenetic Variance in Dopamine D2 Receptor: A Marker of IQ Malleability?" *Translational Psychiatry* 8, no. 1 [2018]: 169); mitochondria (D. C. Geary, "Efficiency of Mitochondrial Functioning as the Fundamental Biological Mechanism of General Intelligence [*g*]," *Psychological Review* 125, no. 6 [2018]: 1028–1050); and brain networks (A. K. Barbey, "Network Neuroscience Theory of Human Intelligence," *Trends in Cognitive Sciences* 22, no. 1 [2018]: 8–20). I suspect *g* is a statistical artifact rather than a fundamental brain mechanism. Intelligence is explained by multiple mental mechanisms that result from neural mechanisms I describe in *Brain-Mind: From Neurons to Consciousness and Creativity* (New York: Oxford University Press, 2019).

R. I. Dunbar advocates the social brain hypothesis in "The Social Brain Hypothesis and Its Implications for Social Evolution," *Annals of Human Biology* 36 (2009): 562–572.

The remark about beer comes from D. Barry, *Dave Barry Turns 40* (New York: Ballantine, 1990).

The following are starter references for the twelve features of intelligence. Problem solving: A. Newell and H. A. Simon, *Human Problem Solving* (Englewood Cliffs, NJ: Prentice Hall, 1972). Planning: G. A. Miller, E. Galanter, and K. Pribram, *Plans and the Structure of Behavior* (New York: Holt, Rinehart and Winston, 1960). Deciding: P. W. Glimcher and E. Fehr, eds., *Neuroeconomics: Decision Making and the Brain*, 2nd ed. (Amsterdam: Academic Press, 2013). Understanding: D. G. Bobrow and A. Collins, eds., *Representation and Understanding: Studies in Cognitive Science* (New York: Academic Press, 1975). Learning: J. H. Holland, K. J. Holyoak, R. E. Nisbett, and P. R. Thagard, *Induction: Processes of Inference, Learning, and Discovery* (Cambridge, MA: MIT Press, 1986). Abstracting: Z. Reznikova, *Animal Intelligence: From Individual to Social Cognition* (Cambridge: Cambridge University Press, 2007). Creating: M. Boden, *The Creative Mind: Myths and Mechanisms*, 2nd ed. (London: Routledge, 2004). Reasoning: J. E. Adler and L. J. Rips, eds., *Reasoning: Studies of Inference and Its Foundations* (Cambridge: Cambridge University Press, 2004). Feeling: A. R. Damasio, *The Feeling of What Happens: Body and Emotion in the Making of Consciousness* (New York: Harcourt Brace, 1999). Communicating: P. Thagard, *Mind-Society: From Brains to Social Sciences and Professions* (New York: Oxford University Press, 2019). Acting: T. D. Lee, *Motor Control in Everyday Actions* (Champaign, IL: Human Kinetics, 2011).

The following are starter references for the eight mechanisms of intelligence. Images: S. M. Kosslyn, W. L. Thompson, and G. Ganis, *The Case for Mental Imagery* (New York: Oxford University Press, 2006). Concepts: P. Blouw, E. Solodkin, P. Thagard, and C. Eliasmith, "Concepts as Semantic Pointers: A Framework and Computational Model," *Cognitive Science* 40 (2016): 1128–1162. Rules: J. R. Anderson, *Rules of the Mind* (Hillsdale, NJ: Erlbaum, 1993). Analogies: K. J. Holyoak and R. G. Morrison, eds., *The Oxford Handbook of Thinking and Reasoning* (New York: Oxford University Press, 2012). Emotions: L. F. Barrett, M. Lewis, and M. Haviland-Jones, eds., *Handbook of Emotions*, 4th ed. (New York: Guilford, 2016). Language: R. Jackendoff, *Foundations of Language: Brain, Meaning, Grammar, Evolution* (Oxford: Oxford University Press, 2002). Intentional actions: T. Schröder, T. C. Stewart, and P. Thagard, "Intention, Emotion, and Action: A Neural Theory Based on Semantic Pointers," *Cognitive Science* 38 (2014): 851–880. Consciousness: Stanislas Dehaene, *Consciousness and the Brain: Deciphering How the Brain Codes Our Thoughts* (New York: Viking, 2014). Many more references and arguments could be given to justify assigning these mechanisms to people in accord with the Attribution Procedure.

The lasagna analogy comes from *CNN Tonight*, February 20, 2019, https://www.cnn.com/videos/politics/2019/02/20/lasagna-of-lies-russia-mueller-probe-don-lemon-chris-cuomo-handoff-sot-ctn-vpx.cnn. The integrative neural theory of emotion is described in Thagard, *Brain-Mind*; and I. Kajić, T. Schröder, T. C. Stewart, and P. Thagard, "The Semantic Pointer Theory of Emotions," *Cognitive Systems Research* 58 (2019): 35–53.

J.-M. Fellous and M. A. Arbib consider the functions of emotions in *Who Needs Emotions? The Brain Meets the Robot* (Oxford: Oxford University Press, 2005). T. E. Feinberg and J. M. Mallatt discuss the adaptive advantages of consciousness in *The Ancient Origins of Consciousness: How the Brain Created Experience* (Cambridge, MA: MIT Press, 2018).

D. C. Kidd and E. Castano find that reading literary fiction improves understanding of other people's minds in "Reading Literary Fiction Improves Theory of Mind," *Science* 342, no. 6156 (2013): 377–380.

Chapter 3: Marvelous Machines

The paper that first interested me in AI was M. Minsky's "A Framework of Representing Knowledge," in *The Psychology of Computer Vision* (New York: McGraw-Hill, 1975). Histories of AI are provided in Boden, *Mind as Machine*; and N. J. Nilsson, *The Quest for Artificial Intelligence: A History of Ideas and Achievements* (Cambridge: Cambridge University Press, 2010). Recent developments are described in M. Ford, ed., *Architects of Intelligence: The Truth about AI from the People Building It* (Birmingham, UK: Packt, 2018); S. Gerrish, *How Smart Machines Think* (Cambridge, MA: MIT Press, 2018); and T. Walsh, *Machines That Think: The Future of Artificial Intelligence* (Amherst, NY: Prometheus Books, 2018). N. Brown and T. Sandholm report on "Superhuman AI for Multiplayer Poker," *Science* 365, no. 6456 (2019): 885–890. M. Mitchell's *Artificial Intelligence: A Guide for Thinking Humans* (New York: Farrar, Straus and Giroux, 2019) provides a good introduction to the state of AI today.

Leading institutions focused on machine intelligence include the Allen Institute for Artificial Intelligence (Seattle), DeepMind (London), Google Brain (Mountain View), Mila (Montreal), OpenAI (San Francisco), the Turing Institute (London), and the Vector Institute (Toronto). My favorite AI researchers in universities include Yoshua Bengio (Montreal), Chris Eliasmith (Waterloo), Ken Forbus (Northwestern), Ashok Goel (Georgia Tech), Geoffrey Hinton (Toronto), John Laird (Michigan), Patrick Langley (ISRE), and Judea Pearl (UCLA). Project: prepare report cards on how well their programs satisfy my twenty benchmarks. I. Kotseruba and J. K. Tsotsos provide a comprehensive comparison of cognitive architectures in "40 Years of Cognitive Architectures: Core Cognitive Abilities and Practical Applications," *Artificial Intelligence Review* 53, no. 1 (2020): 17–94.

How IBM Watson won at *Jeopardy!* is described in D. A. Ferrucci, "Introduction to 'This Is Watson,'" *IBM Journal of Research and Development* 56, nos. 3–4 (2012): 1:1–1:15; and in other articles in the May–June 2012 issue of the *IBM Journal of Research and Development*. For later applications, see Y. Chen, J. D. Elenee Argentinis, and G. Weber, "IBM Watson: How Cognitive Computing Can Be Applied to Big Data Challenges in Life Sciences Research," *Clinical Therapeutics* 38, no. 4 (2016): 688–701; J. G. Hamilton, Ma. G. Garzon, J. S. Westerman, E. Shuk, J. L. Hay, C. Walters, E. Elkin, et al., "'A Tool, Not a Crutch': Patient Perspectives about IBM Watson for Oncology Trained by Memorial Sloan Kettering," *Journal of Oncology Practice* 15 (2019): e277–e288; L. R. Varshney, F. Pinel, K. R. Varshney, D. Bhattacharjya, A. Schorgendorfer, and Y. M. Chee, "A Big Data Approach to Computational Creativity: The Curious Case of Chef Watson," *IBM Journal of Research and Development* 63, no. 1 (2019): 7:1–7:18. See also the IBM web pages on Watson and Project Debater.

Analogy could be added to Watson using techniques from case-based reasoning (D. B. Leake, *Case-Based Reasoning: Experiences, Lessons, and Future Directions* [Menlo Park, CA: AAAI Press/MIT Press, 1996]) and structure mapping (K. D. Forbus, R. W. Ferguson, A. Lovett, and D. Gentner, "Extending SME to Handle Larger-Scale Cognitive Modeling," *Cognitive Science* 41 [2017]: 1152–1201). In chapter 6 of *Brain-Mind: From Neurons to Consciousness to Creativity* (New York: Oxford University Press, 2019), I describe mental and neural mechanisms for analogy that have yet to be modeled computationally but could be accommodated in the semantic pointer architecture of C. Eliasmith; see his *How to Build a Brain: A Neural Architecture for Biological Cognition* (Oxford: Oxford University Press, 2013).

Deep learning is explained in I. Goodfellow, Y. Bengio, and A. Courville, *Deep Learning* (Cambridge, MA: MIT Press, 2016); Y. LeCun, Y. Bengio, and G. Hinton, "Deep Learning," *Nature* 521, no. 7553 (2015): 436–444. E. J. Topol, in both "High-Performance Medicine: The Convergence of Human and Artificial Intelligence," *Nature Medicine* 25, no. 1 (2019): 44–56, and *Deep Medicine: How Artificial Intelligence Can Make Healthcare Human Again* (New York: Basic Books, 2019), reviews applications of deep learning in medicine, as do A. Esteva, A. Robicquet, B. Ramsundar, V. Kuleshov, M. DePristo, K. Chou, C. Cui, et al., in "A Guide to Deep Learning in Healthcare," *Nature*

Medicine 25, no. 1 (2019): 24–29. Reinforcement learning is reviewed in R. S. Sutton and A. G. Barto, *Reinforcement Learning: An Introduction* (Cambridge, MA: MIT Press, 2018).

DeepMind's successes with AlphaZero and other programs are described in D. Silver, T. Hubert, J. Schrittwieser, I. Antonoglou, M. Lai, A. Guez, M. Lanctot, et al., "A General Reinforcement Learning Algorithm That Masters Chess, Shogi, and Go through Self-Play," *Science* 382, no. 6419 (2018): 1140–1144; and M. Botvinick, S. Ritter, J. X. Wang, Z. Kurth-Nelson, C. Blundell, and D. Hassabis, "Reinforcement Learning: Fast and Slow," *Trends in Cognitive Science* 23 (2019): 408–422. G. Marcus, "Innateness, AlphaZero, and Artificial Intelligence," *arXiv preprint arXiv:1801.05667*, 2018, provides a critique; see also G. Marcus and E. Davis, *Rebooting AI: Building Artificial Intelligence We Can Trust* (New York: Pantheon, 2019).

Self-driving cars are reviewed in M. Paden, M. Čáp, S. Zheng Yong, D. Yershov, and E. Frazzoli, "A Survey of Motion Planning and Control Techniques for Self-Driving Urban Vehicles," *IEEE Transactions on Intelligent Vehicles* 1, no. 1 (2016): 33–55; W. Schwarting, J. Alanso-Mora, and D. Rus, "Planning and Decision-Making for Autonomous Vehicles," *Annual Review of Control, Robotics, and Autonomous Systems* 1 (2018): 187–210. For an accessible guide, see A. Davies, "The *Wired* Guide to Self-Driving Cars," *Wired*, December 13, 2018, https://www.wired.com/story/guide-self-driving-cars. Current limitations are described in N. E. Boudette, "Despite High Hopes, Self-Driving Cars Are 'Way in the Future,'" *New York Times*, July 17, 2019, https://www.nytimes.com/2019/07/17/business/self-driving-autonomous-cars.html?smid=nytcore-ios-share.

M. B. Hoy, "Alexa, Siri, Cortana, and More: An Introduction to Voice Assistants," *Medical Reference Services Quarterly* 37, no. 1 (2018): 81–88, reviews virtual assistants. How Alexa works is described in R. Baguley and C. McDonald, "Appliance Science," CNET, August 4, 2016, https://www.cnet.com/news/appliance-science-alexa-how-does-alexa-work-the-science-of-amazons-echo. How Siri works is described in A. Goel, "How Does Siri Work?" *Magoosh Data Science Blog*, February 2, 2018, https://magoosh.com/data-science/siri-work-science-behind-siri.

Google Translate is described in Y. Wu, M. Schuster, Z. Chen, Q. V. Le, M. Norouzi, W. Macherey, M. Krikun, et al., "Google's Neural Machine Translation System: Bridging the Gap between Human and Machine Translation," *arXiv preprint arXiv:1609.08144*, 2016. D. Hofstadter provides a critique in https://www.theatlantic.com/technology/archive/2018/01/the-shallowness-of-google-translate/551570/. In 2020, OpenAI announced GPT-3 which has generated some impressive texts but still lacks the word-to-world semantics essential for natural language understanding. For philosophical examination, see https://dailynous.com/2020/07/30/philosophers-gpt-3/.

Netflix's recommender system is described in Gerrish, *How Smart Machines Think*; and C. A. Gomez-Uribe and Neil Hunt, "The Netflix Recommender System: Algorithms, Business Value, and Innovation," *ACM Transactions on Management Information Systems (TMIS)* 6, no. 4 (2016): 13. See also L. Plummer, "This Is How Netflix's Top-Secret Recommendation System Works," *Wired*, August 22, 2017, https://www.wired.co.uk/article/how-do-netflixs-algorithms-work-machine-learning-helps-to-predict-what-viewers-will-like. Amazon's algorithms are sketched in B. Smith and G. Linden, "Two Decades of Recommender Systems at Amazon.com," *IEEE Internet Computing* 21 (May–June 2017): 12–18, https://www.computer.org/csdl/mags/ic/2017/03/mic2017030012.html.

Predictions about when AI might catch up to human intelligence are in Ford, *Architects of Intelligence*, 527; K. Grace, J. Salvatier, A. Dafoe, B. Zhang, and O. Evans, "When Will AI Ever Exceed Human Performance? Evidence from AI Experts," *Journal of Artificial Intelligence Research* 62 (2018): 729–754; https://aiimpacts.org; "Experts Predict When Artificial Intelligence Will Exceed Human Performance," *MIT Technology Review*, May 31, 2017, https://www.technologyreview.com/s/607970/experts-predict-when-artificial-intelligence-will-exceed-human-performance; and S. Johnson, "Human-like A.I. Will Emerge in 5 to 10 Years, Say Experts," *Big Think*, September 26, 2018, https://bigthink.com/surprising-science/computers-smart-as-humans-5-years. The fact that experts totally disagree shows that nobody knows when (or if) artificial general intelligence will occur.

It might seem that a good strategy for elevating machine intelligence would be to amalgamate different programs. Could a hybrid somehow incorporate IBM Watson with DeepMind's deep reinforcement learning and Waymo's sensory-motor systems? The creation of hybrids and even the arrangement of communication between these different kinds of programs are highly problematic. In general, it should seem that computers could communicate with each other quickly simply by transferring files, but that transfer presumes that the files are in the same formats or easily translatable. The linguistic representations used by IBM Watson are starkly different from the neural network representations used by DeepMind and the Bayesian networks used by some self-driving cars. So hybrids are not an easy solution to the weaknesses of particular machines.

Chapter 4: Amazing Animals

General sources on animal intelligence include De Waal, *Are We Smart Enough*; Z. Reznikova, *Animal Intelligence: From Individual to Social Cognition* (Cambridge: Cambridge University Press, 2007); C. D. L. Wynne, *Do Animals Think?* (Princeton, NJ: Princeton University Press, 2004); and T. R. Zentall and E. A. Wasserman, eds., *The Oxford Handbook of Comparative Cognition* (Oxford: Oxford University Press, 2012). L. Chittka and K. Jensen compare concepts across species in "Animal Cognition: Concepts from Apes to Bees," *Current Biology* 21, no. 3 (2011): R116–R119. C. M. Sanz, J. Call, and C. Boesch review tool use in animals in *Tool Use in Animals: Cognition and Ecology* (Cambridge: Cambridge University Press, 2013). Animal emotions are discussed in F. B. M. de Waal, *Mama's Last Hug: Animals and Human Emotions* (New York: Norton, 2019). K. Andrews, *How to Study Animal Minds* (Abingdon: Routledge, 2020), examines the methods of comparative psychology.

On bee cognition, see L. Chittka, "Bee Cognition," *Current Biology* 27, no. 19 (2017): R1049–R1053; J. L. Gould, "Honey Bee Cognition," *Cognition* 37 (1990): 83–103; T. Hanson, *Buzz: The Nature and Necessity of Bees* (New York: Basic Books, 2018); and R. Menzel and M. Giurfa, "Dimensions of Cognition in an Insect, the Honeybee," *Behavioral Cognitive Neuroscience Review* 5, no. 1 (2006): 24–40. For a review of honeybee vision, see A. Avarguès-Weber, D. d'Amaro, M. Metzler, V. Finke, D. Baracchi, and A. G. Dyer, "Does Holistic Processing Require a Large Brain? Insights from Honeybees and Wasps in Fine Visual Recognition Tasks," *Frontiers in Psychology* 9 (2018): 1313; and A. Avarguès-Weber, T. Mota, and M. Giurfa, "New Vistas on Honey Bee Vision," *Apidologie* 43, no. 3 (2012): 244–268. A. Avarguès-Weber, A. G. Dyer, M. Combe, and M. Giurfa, "Simultaneous Mastering of Two Abstract Concepts by the Miniature Brain of Beas," *Proceedings of the National Academy of Sciences* 109 (2012): 7481–7486, describe abstract relations (quote from p. 7481). S. Alem, C. J. Perry, X. Zhu, O. J. Loukola, T. Ingraham, E. Sovik, and L. Chittka describe social learning by bumblebees in "Associative Mechanisms Allow for Social Learning and Cultural Transmission of String Pulling in an Insect," *PLOS Biology* 14, no. 10 (2016): e1002564. On bee imagery, contrast Chittka, "Bee Cognition," R1052, with Gould, "Honey Bee Cognition," 91; Z. Reznikova, *Animal Intelligence: From Individual to Social Cognition* (Cambridge: Cambridge University Press, 2007), 165; and T. E. Feinberg and J. M. Mallatt, *The Ancient Origins of Consciousness: How the Brain Created Experience* (Cambridge, MA: MIT Press, 2016), 184. H. Ai, K. Kai, A. Kumaraswamy, H. Ikeno, and T. Wachtler analyze the neural basis of waggle dances in "Interneurons in the Honeybee Primary Auditory Center Responding to Waggle Dance–Like Vibration Pulses," *Journal of Neuroscience* 37, no. 44 (2017): 10624–10635. On emotion-like states in bees, see D. Baracchi, M. Lihoreau, and M. Giurfa, "Do Insects Have Emotions? Some Insights from Bumble Bees," *Frontiers in Behavioral Neuroscience* 11 (2017): 157; M. Bateson, S. Desire, S. E. Gartside, and G. A. Wright, "Agitated Honeybees Exhibit Pessimistic Cognitive Biases," *Current Biology* 21, no. 12 (2011): 1070–1073; and Clint J. Perry, Andrew B. Barron, and Lars Chittka, "The Frontiers of Insect Cognition," *Current Opinion in Behavioral Sciences* 16 (2017): 111–118. A. B. Barron and C. Klein argue that insects are capable of consciousness in "What Insects Can Tell Us about the Origins of Consciousness," *Proceedings of the National Academy of Sciences* 113, no. 18 (2016): 4900–4908. J. Groening, D. Venini, and M. V. Srinivivasan report that bees do not increase morphine in response to damage in "In Search of Evidence for the Experience of Pain in Honeybees: A Self-Administration Study," *Scientific Reports* 7 (2017): 45825. C. Solvi, S. Gutierrez Al-Khudhairy, and L. Chittka describe cross-modal object recognition in bumblebees in "Bumble Bees Display Cross-Modal Object Recognition between Visual and Tactile Senses," *Science* 367, no. 6480 (2020): 910–912. Project: use the brain's grid cells to consider whether bees can have cognitive maps without imagery.

On octopus cognition, see A.-S. Darmaillacq, L. Dickel, and J. Mather, *Cephalopod Cognition* (Cambridge: Cambridge University Press, 2014); P. Godfrey-Smith, *Other Minds: The Octopus, the Sea, and the Deep Origins of Consciousness* (New York: Farrar, Straus and Giroux, 2016); B. Hochner, "An Embodied View of Octopus Neurobiology," *Current Biology* 22, no. 20 (2012): R887–R892; and J. A. Mather and L. Dickel, "Cephalopod Complex Cognition," *Current Opinion in Behavioral Sciences* 16 (2017): 131–137. E. Edsinger and G. Dölen describe octopuses on MDMA becoming more sociable in "A Conserved Role for Serotonergic Neurotransmission in Mediating Social Behavior in Octopus," *Current Biology* 28, no. 19 (2018): 3136–3142. R. J. Crook, R. T. Hanlon, and E. T. Walters report nociceptors in squid in

"Squid Have Nociceptors That Display Widespread Long-Term Sensitization and Spontaneous Activity after Bodily Injury," *Journal of Neuroscience* 33, no. 24 (2013): 10021–10026; and P. L. R. Andrews, A.-S. Darmaillacq, N. Dennison, I. G. Gleadall, P. Hawkins, J. B. Messenger, Daniel Osorio, et al., discuss evidence for pain in cephalopods in "The Identification and Management of Pain, Suffering and Distress in Cephalopods, including Anaesthesia, Analgesia and Humane Killing," *Journal of Experimental Marine Biology* 447 (2013): 46–64. D. Scheel, S. Chancellor, M. Hing, M. Lawrence, S. Linquist, and P. Godfrey-Smith describe octopus communities in "A Second Site Occupied by *Octopus tetricus* at High Densities, with Notes on Their Ecology and Behavior," *Marine and Freshwater Behaviour and Physiology* 50, no. 4 (2017): 285–291.

General works on bird cognition include T. Birkhead, *Bird Sense: What It's Like to Be a Bird* (London: Bloomsbury, 2012); N. Emery, *Bird Brain: An Exploration of Avian Intelligence* (Princeton, NJ: Princeton University Press, 2016); O. Güntürkün and T. Bugnyar, "Cognition without Cortex," *Trends in Cognitive Sciences* 20, no. 4 (2016): 291–303; S. Olkowicz, M. Kocourek, R. K. Lucan, M. Portes, W. T. Fitch, S. Herculano-Houzel, and P. Nemec, "Birds Have Primate-Like Numbers of Neurons in the Forebrain," *Proceedings of the National Academy of Sciences* 113, no. 26 (2016): 7255–7260; and C. ten Cate and S. D. Healy, *Avian Cognition* (Cambridge: Cambridge University Press, 2017). M. R. Papini, J. C. Penagos-Corzo, and A. M. Pérez-Acosta discuss bird emotions in "Avian Emotions: Comparative Perspectives on Fear and Frustration," *Frontiers in Psychology* 9 (2018): 2707. On ravens and crows, see R. Gruber, M. Schiestl, M. Boeckle, A. Frohnwieser, R. Miller, R. D. Gray, N. S. Clayton, et al., "New Caledonian Crows Use Mental Representations to Solve Metatool Problems," *Current Biology* 29, no. 4 (2017): 686–692 e3; B. Heinrich, *Mind of the Raven* (New York: HarperCollins, 2006) (quote from p. 191); B. Heinrich and T. Bugnyar, "Testing Problem Solving in Ravens: String-Pulling to Reach Food," *Ethology* 111, no. 10 (2005): 962–976; I. F. Jacobs, A. von Bayern, G. Martin-Ordas, L. Rat-Fischer, and M. Osvath, "Corvids Create Novel Causal Interventions After All," *Proceedings of the Royal Society B* 282, no. 1806 (2015): 20142504; C. Kabadayi and M. Osvath, "Ravens Parallel Great Apes in Flexible Planning for Tool-Use and Bartering," *Science* 357, no. 6347 (2017): 202–204; S. A. Jelbert, R. J. Hosking, A. H. Taylor, and R. D. Gray, "Mental Template Matching Is a Potential Cultural Transmission Mechanism for New Caledonian Crow Tool Manufacturing

Traditions," *Scientific Reports* 8, no. 1 (2018): 8956; J. Marzluff and T. Angell, *Gifts of the Crow* (New York: Atria, 2012); J. J. A. Müller, J. J. M. Massen, T. Bugnyar, and M. Osvath, "Ravens Remember the Nature of a Single Reciprocal Interaction Sequence over 2 Days and Even after a Month," *Animal Behaviour* 128 (2017): 69–78; A. M. P. von Bayern, S. Danel, A. M. I. Auersperg, B. Mioduszewska, and A. Kacelnik, "Compound Tool Construction by New Caledonian Crows," *Scientific Reports* 8, no. 1 (2018): 15676; and A. A. Wright, J. F. Magnotti, J. S. Katz, K. Leonard, A. Vernouillet, and D. M. Kelly, "Corvids Outperform Pigeons and Primates in Learning a Basic Concept," *Psychological Science* 28 (2017): 437–444.

Good sources on dog cognition include M. K. Bensky, S. D. Gosling, and D. L. Sinn, "The World from a Dog's Point of View," *Advances in the Study of Behavior* 45 (2013): 209–406; G. Berns, *What It's Like to Be a Dog, and Other Adventures in Animal Neuroscience* (New York: Basic Books, 2017); B. Hare and V. Woods, *The Genius of Dogs* (New York: Dutton, 2013); and especially Á. Miklósi, *Dog Behavior, Evolution, and Cognition* (Oxford: Oxford University Press, 2015). G. S. Berns, A. M. Brooks, and M. Spivak study dog sense of smell in "Scent of the Familiar: An fMRI Study of Canine Brain Responses to Familiar and Unfamiliar Human and Dog Odors," *Behavioral Processes* 110 (2015): 37–46. A. Prichard, R. Chhibber, K. Athanassiades, M. Spivak, and G. S. Berns examine fast neural learning in "Fast Neural Learning in Dogs: A Multimodal Sensory fMRI Study," *Scientific Reports* 8, no. 1 (2018): 14614. C. Fugazza, A. Moesta, A. Pogány, and A. Miklósi study social learning in "Social Learning from Conspecifics and Humans in Dog Puppies," *Scientific Reports* 8, no. 1 (2018): 9257. M. V. Kujala reviews dog emotions in "Canine Emotions as Seen through Human Social Cognition," *Animal Sentience* 14, no. 1 (2017). A. Horowitz argues that a guilty look is not guilt in "Disambiguating the 'Guilty Look': Salient Prompts to Familiar Dog Behavior," *Behavioral Processes* 81, no. 3 (2009): 447–452. A. Horowitz found that dogs recognize their own urine in "Smelling Themselves: Dogs Investigate Their Own Odours Longer When Modified in an 'Olfactory Mirror' Test," *Behavioral Processes* 143 (2017): 17–24. See chapter 7 for a discussion of whether dogs are jealous.

Dolphin cognition is reviewed in J. Gregg, *Are Dolphins Really Smart? The Mammal behind the Myth* (Oxford: Oxford University Press, 2013) (language evaluation, 136); L. M. Herman, "What Laboratory Research Has Told Us about Dolphin Cognition," *International Journal of Comparative Psychology* 23 (2010): 310–330; D. L. Herzing and C. M. Johnson, *Dolphin Communication and Cognition: Past, Present, and Future* (Cambridge, MA: MIT Press, 2015); E. Mercado and C. M. DeLong, "Dolphin Cognition: Representations and Processes in Memory and Perception," *International Journal of Comparative Psychology* 23, no. 3 (2010). R. J. Schusterman, J. A. Thomas, and F. Glenn Wood survey dolphin communication in *Dolphin Cognition and Behavior: A Comparative Approach* (Hillsdale, NJ: Lawrence Erlbaum, 1986). M. Bearzi and C. B. Stanford compare dolphins and apes in *Beautiful Minds: The Parallel Lives of Great Apes and Dolphins* (Cambridge, MA: Harvard University Press, 2008). R. Morrison and D. Reiss describe mirror self-recognition by dolphins in "Precocious Development of Self-Awareness in Dolphins," *PLOS One* 13, no. 1 (2018): e0189813. H. S. Mortenson, B. Pakkenberg, M. Dam, R. Dietz, C. Sonne, B. Mikkelsen, and N. Eriksen analyze dolphin brains in "Quantitative Relationships in Delphinid Neocortex," *Frontiers in Neuroanatomy* 8 (2014): 132. Tail walking learning is reported in E. Young, "A Once-Captive Dolphin Has Introduced Her Friends to a Silly Trend," *Atlantic*, September 5, 2018, https://www.theatlantic.com/science/archive/2018/09/dolphins-tail-walking-trend/569314.

Chimpanzee cognition is covered in E. Lonsdorf, S. R. Ross, and T. Matsuzawa, *The Mind of the Chimpanzee: Ecological and Experimental Perspectives* (Chicago: University of Chicago Press, 2010). M. Tomasello and J. Call review primate cognition in *Primate Cognition* (New York: Oxford University Press, 1997). E. Herrmann, B. Hare, J. Call, and M. Tomasello ("Differences in the Cognitive Skills of Bonobos and Chimpanzees," *PLOS One* 5, no. 8 [2010]: e12438) and J. K. Rilling, J. Scholz, T. M. Preuss, M. F. Glasser, B. K. Errangi, and T. E. Behrens ("Differences between Chimpanzees and Bonobos in Neural Systems Supporting Social Cognition," *Social Cognitive Affective Neuroscience* 7, no. 4 [2012]: 369–379) compare chimpanzees and bonobos. Dorothy L. Cheney and Robert M. Seyfarth compare chimpanzees with baboons and humans in *Baboon Metaphysics: The Evolution of a Social Mind* (Chicago: University of Chicago Press, 2008). Bonobos are described in F. B. M. de Waal, *The Bonobo and the Atheist: In Search of Humanism among the Primates* (New York: Norton, 2013). W. Köhler describes insight learning in *The Mentality of Apes* (London: Kegan Paul, Trench, Taubner, 1927), although some ethologists still prefer the alternative explanation of trial-and-error learning. A. Whiten compares social learning in child and chimpanzee in "Social Learning and Culture in Child and Chimpanzee," *Annual Review of Psychology* 68 (2017): 129–154. S. Musgrave, D. Morgan, E.

Lonsdorf, R. Mundry, and C. Sanz report teaching of tool use by chimpanzees in "Tool Transfers Are a Form of Teaching among Chimpanzees," *Scientific Reports* 6 (2016): 34783. D. M. Altschul, E. K. Wallace, R. Sonnweber, M. Tomonaga, and A. Weiss describe chimpanzee performance on touch screen tasks in "Chimpanzee Intellect: Personality, Performance and Motivation with Touchscreen Tasks," *Royal Society Open Science* 4, no. 5 (2017): 170169. M. E. Kret, A. Muramatsu, and T. Matsuzawa compare emotions in humans and chimpanzees in "Emotion Processing across and within Species: A Comparison between Humans (*Homo sapiens*) and Chimpanzees (*Pan troglodytes*)," *Journal of Comparative Psychology* 132, no. 4 (2018): 395–409. W. T. Finch, *The Evolution of Language* (Cambridge: Cambridge University Press, 2010), describes the limits of nonhuman language.

The sources for table 4.1 include S. Herculano-Houzel, *The Human Advantage: A New Understanding of How Our Brain Became Remarkable* (Cambridge, MA: MIT Press, 2016); S. Herculano-Houzel, "Numbers of Neurons as Biological Correlates of Cognitive Capability," *Current Opinion in Behavioral Sciences* 16 (2017): 1–7; D. Jardim-Messeder, K. Lambert, S. Noctor, F. M. Pestana, M. E. de Castro Leal, M. F. Bertelsen, A. N. Alagaili, et al., "Dogs Have the Most Neurons, Though Not the Largest Brain: Trade-Off between Body Mass and Number of Neurons in the Cerebral Cortex of Large Carnivoran Species," *Frontiers in Neuroanatomy* 11 (2017): 118; S. Olkowicz et al., "Birds Have Primate-Like Numbers of Neurons in the Forebrain," 7255–7260; and various websites including Wikipedia, https://en.wikipedia.org/wiki/List_of_animals_by_number_of_neurons and https://faculty.washington.edu/chudler/facts.html.

Chapter 5: Human Advantages

My list of human advantages expands on M. C. Corballis, *The Recursive Mind: The Origins of Human Language, Thought, and Civilization* (Princeton, NJ: Princeton University Press, 2011); Joseph Henrich, *The Secret of Our Success* (Princeton, NJ: Princeton University Press, 2016); Herculano-Houzel, *The Human Advantage*; D. C. Penn, K. J. Holyoak, and D. J. Povinelli, "Darwin's Mistake: Explaining the Discontinuity between Human and Nonhuman Minds," *Behavioral and Brain Sciences* 31, no. 2 (2008): 109–130; T. Suddendorf, *The Gap: The Science of What Separates Us from Other Animals* (New York: Basic Books, 2013); and Wynne, *Do Animals Think?*. This list is open to revision based on ongoing research. A few animals such as jays may be able to think about past, future, and the minds of others to a limited degree: N. S. Clayton, T. J. Bussey, and A. Dickenson, "Can Animals Recall the Past and Plan for the Future?" *Nature Reviews Neuroscience* 4, no. 8 (2003): 685. Humans can think about their own minds and the minds of others even before and after death. Cr. Krupenye, F. Kano, S. Hirata, J. Call, and M. Tomasello claim, in "Great Apes Anticipate That Other Individuals Will Act According to False Beliefs," *Science* 354, no. 6308 (2016): 110–114, that great apes implicitly understand false beliefs, suggesting that animals can hypothesize hidden causes when they attribute mental states; but it is more plausible that animals perform multimodal rule simulation (Ashley Keefner, "Corvids Infer the Minds of Conspecifics," *Biology and Philosophy* 31, no. 2 [2015]: 267–281). A. H. Taylor, R. Miller, and R. D. Gray contend that New Caledonian crows can infer hidden causal agents in "New Caledonian Crows Reason about Hidden Causal Agents," *Proceedings of the National Academy of Sciences* 109, no. 40 (2012): 16389–16391. A. H. Taylor, in "Corvid Cognition," *WIREs Cognitive Science* 5, no. 3 (2014): 361–372, describes crows' use of "meta-tools," or using tools to retrieve tools, which is not as creative as making tools that make tools.

AI research on learning how to learn is beginning: E. Real, C. Liang, D. R. So, and Quoc V. Le develop novel techniques for machine learning of machine learning algorithms in "AutoML-Zero: Evolving Machine Learning Algorithms from Scratch," *arXiv preprint arXiv:2003.03384*.

The hypothesis that cooking played a major role in the evolution of large human brains is advocated in Herculano-Houzel, *The Human Advantage*; C. Organ, C. L. Nunn, Z. Machanda, and R. W. Wrangham, "Phylogenetic Rate Shifts in Feeding Time during the Evolution of *Homo*," *Proceedings of the National Academy of Sciences* 108, no. 35 (2011): 14555–14559; and R. W. Wrangham, *Catching Fire: How Cooking Made Us Human* (New York: Basic Books, 2009). The major problem with this hypothesis is lack of archaeological evidence for cooking 1.5 million years ago, but an African hearth around a million years old has been found: K. Miller, "Archaeologists Find Earliest Evidence of Humans Cooking with Fire," *Discover*, December 16, 2013, https://www.discovermagazine.com/the-sciences/archaeologists-find-earliest-evidence-of-humans-cooking-with-fire. Alternative (or complementary) hypotheses for the evolution of large brains are tool use (D. M. Rumbaugh, M. J. Beran, and W. A. Hillix, "Cause-Effect Reasoning in Humans and Animals," in *The*

Evolution of Cognition, ed. C. Heyes and L. Huber [Cambridge, MA: MIT Press, 2000]); social emotions (E. Jablonka, S. Ginsburg, and D. Dor, "The Co-evolution of Language and Emotions," *Philosophical Transactions of the Royal Society B: Biological Sciences* 367, no. 1599 [2012]: 2152–2159); and social learning (M. Muthukrishna, M. Doebeli, M. Chudek, and J. Henrich, "The Cultural Brain Hypothesis: How Culture Drives Brain Expansion, Sociality, and Life History," *PLOS Computational Biology* 14, no. 11 [2018]: e1006504).

Corballis, *The Recursive Mind*, and T. Suddendorf, *The Gap: The Science of What Separates Us from Other Animals* (New York: Basic Books, 2013), explain recursive thinking. I take the term "looping effect" from I. Hacking, who applies it to how social concepts can change society in *The Social Construction of What?* (Cambridge, MA: Harvard University Press, 1999). Hofstadter, *I Am a Strange Loop*, considers persons as "strange loops." I avoid using the popular but misleading terms "theory of mind" and "mental time travel."

You can watch a crow solve an eight-step problem on YouTube, https://www.youtube.com/watch?v=Gui3IswQ0DI. F. S. Medina, A. H. Taylor, G. R. Hunt, and R. D. Gray test crows' responses to mirrors in "New Caledonian Crows' Responses to Mirrors," *Animal Behavior* 82, no. 5 (2011): 981–993.

On human rule-based problem solving, see A. Newell and H. A. Simon, *Human Problem Solving* (Englewood Cliffs, NJ: Prentice Hall, 1972); and J. E. Laird, C. Lebiere, and P. S. Rosenbloom, "A Standard Model of the Mind: Toward a Common Computational Framework across Artificial Intelligence, Cognitive Science, Neuroscience, and Robotics," *AI Magazine* 38, no. 4 (2017): 13–26.

R. C. Berwick and N. Chomsky propose that a mutation led to human language in *Why Only Us: Language and Evolution* (Cambridge, MA: MIT Press, 2016), 70.

On theoretical neuroscience (also known as computational neuroscience), see P. Dayan and L. F. Abbot, *Theoretical Neuroscience: Computational and Mathematical Modeling of Neural Systems* (Cambridge, MA: MIT Press, 2001); and T. J. Sejnowski, *The Deep Learning Revolution* (Cambridge, MA: MIT Press, 2018).

The semantic pointer theory of mind is developed by Eliasmith, *How to Build a Brain*. See also Thagard, *Brain-Mind*; and P. Thagard, *Natural Philosophy: From Social Brains to Knowledge, Reality, Morality, and Beauty* (New York: Oxford University Press, 2019). The semantic pointer theory of emotions is stated in Thagard, *Brain-Mind*; and I. Kajić, T. Schröder, T. C. Stewart, and P. Thagard, "The Semantic Pointer Theory of Emotions," *Cognitive Systems Research* 58 (2019): 35–53. S. D. Kreibig, "Autonomic Nervous System Activity in Emotion: A Review," *Biological Psychology* 84 (2010): 394–421, table 1, displays the physiological similarity of anger and fear.

M. Scheffer, *Critical Transitions in Nature and Society* (Princeton, NJ: Princeton University Press, 2009), describes critical transitions in nature and society; and Thagard, *Natural Philosophy*, analyzes emergence that results from such transitions. Emergent properties belong to a whole but not to its parts and are not just aggregates of the properties of the parts, because they result from interactions of the parts.

The Bengio quote on causality comes from his interview in Ford, *Architects of Intelligence*, 18. On causality, see J. Pearl, *Causality: Models, Reasoning, and Inference* (Cambridge: Cambridge University Press, 2000); J. Pearl and D. Mackenzie, *The Book of Why: The New Science of Cause and Effect* (New York: Basic Books, 2018) (quote on p. 13); D. C. Penn and D. J. Povinelli, "Causal Cognition in Human and Nonhuman Animals: A Comparative, Critical Review," *Annual Review of Psychology* 58 (2007): 97–118; Thagard, *Brain-Mind*; and Thagard, *Natural Philosophy*. The three-analysis of causality comes from Thagard, *Brain-Mind*, 100. An alternative route to causality that tracks sensory-motor-sensory patterns might be to use autoencoders in neural networks combined with model-based reinforcement learning.

R. Baillargeon, R. M. Scott, and L. Bian review children's physical knowledge in "Psychological Reasoning in Infancy," *Annual Review of Psychology* 67 (2016): 159–186. Sanz, Call, and Boesch review animals' use of tools in *Tool Use in Animals*. J. Woodward considers the importance of manipulation and intervention for causality and explanation in *Making Things Happen: A Theory of Causal Explanation* (Oxford: Oxford University Press, 2004). A. Karpathy, G. Toderici, S. Shetty, T. Leung, R. Sukthankar, and L. Fei-Fei, "Large-Scale Video Classification with Convolutional Neural Networks," in *2014 IEEE Conference on Computer Vision and Pattern Recognition* (Columbus: IEEE), 1725–1732 apply deep learning to videos, raising the possibility of computer learning of causal relations.

My quote from Homer's *Iliad* comes from the translation by S. Butler, available at https://www.gutenberg.org/ebooks/2199.

On the contributions of abductive inference, conceptual combination, and analogy to forming hypotheses about hidden causes, see P. Thagard,

Computational Philosophy of Science (Cambridge, MA: MIT Press, 1988); P. Thagard, *The Cognitive Science of Science: Explanation, Discovery, and Conceptual Change* (Cambridge, MA: MIT Press, 2012); and Thagard, *Brain-Mind*.

On the importance of emotion for social intelligence, see D. Goleman, *Emotional Intelligence* (New York: Bantam, 1995); P. Salovey, B. T. Detweiler-Bedell, J. B. Detweiler-Bedell, and J. D. Mayer, "Emotional Intelligence," in *Handbook of Emotions*, ed. M. Lewis, J. M. Haviland-Jones, and L. F. Barrett (New York: Guilford Press, 2008), 533–547; and P. Thagard, *Mind-Society: From Brains to Social Sciences and Professions* (New York: Oxford University Press, 2019), which also describes cognitive and emotional mechanisms that enhance communication.

M. Tomasello, *Becoming Human: A Theory of Ontogeny* (Cambridge, MA: Harvard University Press, 2019), describes the social origins of humanity. On empathy and fairness in animals, see De Waal, *Mama's Last Hug*, and many of his other books. M. J. Beran and W. D. Hopkins describe self-control in chimpanzees in "Self-Control in Chimpanzees Relates to General Intelligence," *Current Biology* 28, no. 4 (2018): 574–579, e3. K. N. Laland, *Darwin's Unfinished Symphony: How Culture Made the Human Mind* (Princeton, NJ: Princeton University Press, 2017), 191, proposes that teaching is a major reason why language evolved.

C. Safina gives examples of teaching by animals in *Beyond Words: What Animals Think and Feel* (London: Macmillan, 2015); but G. Csibra and G. Gergely claim that only humans teach in "Natural Pedagogy as Evolutionary Adaptation," *Philosophical Transactions of the Royal Society B: Biological Sciences* 366, no. 1567 (2011): 1149–1157. M. Wooldridge reviews multiagent systems in artificial intelligence in *An Introduction to Multiagent Systems* (Chichester: John Wiley & Sons, 2002). H. Andersen, X. Shen, Y. H. Eng, D. Rus, and M. H. Ang discuss cooperation among autonomous vehicles in "Connected Cooperative Control of Autonomous Vehicles during Unexpected Road Situations," *Mechanical Engineering* 139, no. 12 (2017): S3–S7. The internet device statistic comes from "Internet of Things (IoT) Connected Devices Installed Base Worldwide from 2015 to 2025," Statista, November 27, 2016, https://www.statista.com/statistics/471264/iot-number-of-connected-devices-worldwide.

C. A. Chapman and M. A. Huffman discuss the motivation for thinking that humans are different in "Why Do We Want to Think Humans Are Different?" *Animal Sentience* 3, no. 23 (2018). C. Darwin, *The Descent of Man, and Selection in Relation to Sex* (Princeton, NJ: Princeton University Press), says that human and animal minds differ only in degree, an idea that I challenge in chapter 7 and in more detail in P. Thagard, "Darwin and the Golden Rule: How to Distinguish Differences of Degree from Differences of Kind Using Mechanisms" (forthcoming).

Chapter 6: When Did Minds Begin?

A. S. Reber claims that bacteria have minds in *The First Minds: Caterpillars, 'Karyotes, and Consciousness* (New York: Oxford University Press, 2019). F. Baluska and A. Reber propose mechanisms for cell consciousness in "Sentience and Consciousness in Single Cells: How the First Minds Emerged in Unicellular Species," *BioEssays* 41, no. 3 (2019): 1800229. J. Jaynes says that consciousness is a cultural development in *The Origin of Consciousness in the Breakdown of the Bicameral Mind*. C. Koch, *The Feeling of Life Itself: Why Consciousness Is Widespread but Can't Be Computed* (Cambridge, MA: MIT Press, 2019), claims that bacteria are conscious based on the information integration theory of consciousness critiqued in Thagard, *Natural Philosophy*. C. Parisien and P. Thagard, in "Robosemantics: How Stanley the Volkswagen Represents the World," *Minds and Machines* 18 (2006): 169–178, argue that self-driving cars refute J. Searle, "Minds, Brains, and Programs," *Behavioral and Brain Sciences* 3 (1980): 417–424. For an approximate time line of evolution, see M. Marshall, "Timeline: The Evolution of Life," *New Scientist*, July 14, 2009, https://www.newscientist.com/article/dn17453-timeline-the-evolution-of-life.

R. Descartes, in *The Philosophical Writings of Descartes*, trans. J. Cottingham et al. (Cambridge: Cambridge University Press, 1985), argued that he could not doubt that he was thinking. The attribution diagrams are based on the explanatory coherence theory of P. Thagard, "Explanatory Coherence," *Behavioral and Brain Sciences* 12 (1989): 435–467; and P. Thagard, *Conceptual Revolutions* (Princeton, NJ: Princeton University Press, 1992).

Defenders of plant intelligence include Anthony Trewavas, *Plant Behavior and Intelligence* (Oxford: Oxford University Press, 2014); and M. Gagliano, J. C. Ryan, and P. Vieira, *The Language of Plants: Science, Philosophy, Literature* (Minneapolis: University of Minneapolis Press, 2017). L. Taiz, D. Alkon, A. Draguhn, A. Murphy, M. Blatt, C. Hawes, G. Thiel, et al., debunk plant consciousness in "Plants Neither Possess Nor Require Consciousness," *Trends in Plant Science* 14 (2019): 677–687.

C. Allen and M. Trestman review philosophical ideas about animal consciousness in "Animal Consciousness," *Stanford Encyclopedia of Philosophy*, https://plato.stanford.edu/entries/consciousness-animal. P. Le Neindre, E. Bernard, A. Boissy, X. Boivin, L. Calandreau, N. Delon, B. Deputte, et al., review scientific research on animal consciousness in "Animal Consciousness," *EFSA Supporting Publications* 14, no. 4 (2017): 1196E. T. E. Feinberg and J. M. Mallatt locate the origins of consciousness in insects, fish, and cephalopods in *The Ancient Origins of Consciousness: How the Brain Created Experience* (Cambridge, MA: MIT Press, 2016). Neural theories of consciousness include Dehaene, *Consciousness and the Brain*; and Thagard, *Brain-Mind*.

C. W. Woo, M. Roy, J. T. Buhle, and T. D. Wager review brain mechanisms for human pain in "Distinct Brain Systems Mediate the Effects of Nociceptive Input and Self-Regulation on Pain," *PLOS Biology* 13, no. 1 (2015): e1002036. Advocates of fish pain include V. Braithwaite, *Do Fish Feel Pain?* (Oxford: Oxford University Press, 2001); and Lynne U. Sneddon, Javier Lopez-Luna, David C. C. Wolfenden, Matthew C. Leach, Ana M. Valentim, Peter J. Steenbergen, Nabila Bardine, et al., "Fish Sentience Denial: Muddying the Waters," *Animal Sentience* 3, no. 21 (2018): 1. Skeptics about fish pain include J. D. Rose, R. Arlinghaus, S. J. Cooke, B. K. Diggles, W. Sawynok, E. D. Stevens, and C. D. L. Wynne, "Can Fish Really Feel Pain?" *Fish and Fisheries* 15 (2012): 97–133; and Brian Key, "Why Fish Do Not Feel Pain," *Animal Sentience* 1, no. 3 (2016). M. Kohda, T. Hotta, T. Takeyama, S. Awata, H. Tanaka, J. Y. Asai, and A. L. Jordan claim that fish pass the mirror test for self-consciousness in "If a Fish Can Pass the Mark Test, What Are the Implications for Consciousness and Self-Awareness Testing in Animals?" *PLOS Biology* 17, no. 2 (2019): e3000021; but F. B. M. de Waal prefers a gradualist approach to self-awareness in "Fish, Mirrors, and a Gradualist Perspective on Self-Awareness," *PLOS Biology* 17, no. 2 (2019): e30000112.

B. J. King makes the case for grief in animals in *How Animals Grieve* (Chicago: University of Chicago Press, 2013). More general works on animal emotions include M. Bekoff, *The Emotional Lives of Animals* (Novato, CA: New World Library, 2007); and De Waal, *Mama's Last Hug*. M. A. L. Reggente, F. Alves, C. Nicolau, L. Freitas, D. Cagnazzi, R. W. Baird, and P. Galli describe grief in dolphins in "Nurturant Behavior toward Dead Conspecifics in Free-Ranging Mammals: New Records for Odontocetes and a General Review," *Journal of Mammalogy* 97, no. 5 (2016): 1428–1434. Horowitz, in "Disambiguating the 'Guilty Look,'" debunks guilt in dogs; but De Waal, in *Mama's*

Last Hug, maintains that dogs can be guilty in cases of violation of dominance rules.

Studies on jealousy in dogs include J. Abdai, C. Bano Terencio, P. Perez Fraga, and Á. Miklósi, "Investigating Jealous Behavior in Dogs," *Scientific Reports* 8, no. 1 (2018): 8911; P. Cook, A. Prichard, M. Spivak, and G. S. Berns, "Jealousy in Dogs? Evidence from Brain Imaging," *Animal Sentience* 3, no. 22 (2018): 1; C. R. Harris and C. Prouvost, "Jealousy in Dogs," *PLOS One* 9, no. 7 (2014): e94597; E. W. Mathes and D. J. Deuger, "Jealousy, a Creation of Human Culture?" *Psychological Reports* 51, no. 2 (1982): 351–354; and P. H. Morris, C. Doe, and E. Godsell, "Secondary Emotions in Non-primate Species? Behavioural Reports and Subjective Claims by Animal Owners," *Cognition and Emotions* 22, no. 1 (2007): 3–20.

Advocates of the idea that animals can think with analogies include R. G. Cook and E. A. Wasserman, "Learning and Transfer of Relational Matching-to-Sample by Pigeons," *Psychonomic Bulletin and Review* 14, no. 6 (2007): 1107–1114; J. Fagot and A. Maugard, "Analogical Reasoning in Baboons (*Papio papio*): Flexible Reencoding of the Source Relation Depending on the Target Relation," *Learning and Behavior* 41, no. 3 (2007): 229–237; T. M. Flemming, R. K. R. Thompson, and J. Fagot, "Baboons, like Humans, Solve Analogy by Categorical Abstraction of Relations," *Animal Cognition* 16, no. 3 (2013): 519–524; D. J. Gillan, D. Premack, and G. Woodruff, "Reasoning in the Chimpanzee: I. Analogical Reasoning," *Journal of Experimental Psychology: Animal Behavior Processes* 7 (1981): 1–17; K. J. Holyoak and P. Thagard, *Mental Leaps: Analogy in Creative Thought* (Cambridge, MA: MIT Press, 1995); T. Obozova, A. Smirnova, Z. Zorina, and E. Wasserman, "Analogical Reasoning in Amazons," *Animal Cognition* 18, no. 6 (2015): 1363–1371; D. Premack, "Why Humans Are Unique: Three Theories," *Perspectives in Psychological Science* 5, no. 1 (2010): 22–32; A. Smirnova, Z. Zorina, T. Obozova, and E. Wasserman, "Crows Spontaneously Exhibit Analogical Reasoning," *Current Biology* 25, no. 2 (2015): 256–260; R. K. R. Thompson and D. L. Oden, "Categorical Perception and Conceptual Judgments by Nonhuman Primates: The Paleological Monkey and the Analogical Ape," *Cognitive Science* 24, no. 3 (2000): 363–396. Psychologists call the task shown in figure 6.1 "relational match to sample." Skeptics about animal analogies include Penn, Holyoak, and Povinelli, "Darwin's Mistake," 109–130; S. Dymond and I. Stewart, "Relational and Analogical Reasoning in Comparative Cognition," *International Journal of Comparative Psychology* 29, no. 1 (2016); and

J. Vonk, "Corvid Cognition: Something to Crow About?" *Current Biology* 25, no. 2 (2015): R69–R71. Scientists claimed that the chimpanzee Sarah did proportional analogies, but her performance was never replicated and is open to alternative explanations.

I. Copi has two additional criteria for evaluating analogical inferences in *Introduction to Logic* (New York: Macmillan, 1982), 365. His account is consistent with that in the leading logic book of the nineteenth century, J. S. Mill, *A System of Logic* (London: Longman, 1970), 365.

M. Tegmark, *Life 3.0: Being Human in the Age of Artificial Intelligence* (New York: Alfred A. Knopf, 2017), argues for substrate independence, a position that philosophers call functionalism, which is usually based on arguments about multiple realizability challenged in W. Bechtel and J. Mundale, "Multiple Realizability Revisited: Linking Cognitive and Neural States," *Philosophy of Science* 66 (1999): 175–207. Feinberg and Mallatt, *The Ancient Origins of Consciousness*, and T. E. Feinberg and J. M. Mallatt, *Demystifying Consciousness: How the Brain Creates Experience* (Cambridge, MA: MIT Press, 2018), defend a biological theory of consciousness. P. Thagard, "Energy Requirements Undermine Substrate Independence and Mind-Body Functionalism" (forthcoming in *Philosophy of Science*), argues that the plausibility of substrate independence is undermined by considering how biological and computational systems use energy. Because real-world information processing depends on energy requirements that are substrate dependent, intelligence is substrate dependent. This argument also challenges claims that human minds can be transferred to computers.

Thagard, *Brain-Mind*, and P. Thagard and T. C. Stewart, "Two Theories of Consciousness: Semantic Pointer Competition vs. Information Integration," *Consciousness and Cognition* 30 (2014): 73–90, explain consciousness as competition among semantic pointers. Another theory is that consciousness occurs when information is broadcast across brain regions: Dehaene, *Consciousness and the Brain*; S. Dehaene, H. Lau, and S. Kouider, "What Is Consciousness, and Could Machines Have It?" *Science* 358, no. 6362 (2017): 486–492. The information integration theory of consciousness (G. Tononi, M. Boly, M. Massimi, and C. Koch, "Integrated Information Theory: From Consciousness to Its Physical Substrate," *Nature Reviews Neuroscience* 17, no. 7 [2016]: 450–461) implies that computers are already conscious, but this theory has numerous mathematical and empirical problems described in Thagard and Stewart, "Two Theories of Consciousness," and Thagard,

Natural Philosophy. J.-M. Fellous and Michael A. Arbib, *Who Needs Emotions? The Brain Meets the Robot* (Oxford: Oxford University Press, 2005), considers robot emotions.

J. H. Lau, T. Cohn, Timothy Baldwin, Julian Brooke, and Adam Hammond, "Deep-Speare: A Joint Neural Model of Poetic Language, Meter, and Rhyme," *arXiv preprint arXiv:1807.03491*, 2018, give a neural network model of poetry. M. Boden, *Creativity and Art: Three Roads to Surprise* (Oxford: Oxford University Press, 2010), and M. du Sautoy, *Creativity Code: Art and Innovation in the Age of AI* (Cambridge, MA: MIT Press, 2019), provide examples of computer creativity, which is challenged in Kelly, "A Philosopher Argues That an AI Can't Be an Artist." Du Sautoy, *Creativity Code*, 283, links creativity to consciousness, but whether creativity requires or is merely enhanced by consciousness (through motivation, evaluation, and reflection) is unclear. Thagard, *Brain-Mind*, discusses the creation of new methods, and Thagard, *Mind-Society*, analyzes norms. W. Reich, *Schoenberg: A Critical Biography*, trans. L. Black (London: Longman, 1971), describes the development of Schoenberg's musical methods.

T. Westby and C. J. Conselice calculate the number of civilizations on other planets in "The Astrobiological Copernican Weak and Strong Limits for Intelligent Life," *Astrophysical Journal* 896, no. 1 (2020): 58. Another skeptical response is N. R. Longrich, "Evolution Tells Us We Might Be the Only Intelligent Life in the Universe," *The Conversation*, October 18, 2019, https://theconversation.com/evolution-tells-us-we-might-be-the-only-intelligent-life-in-the-universe-124706.

Chapter 7: The Morality of Bots and Beasts

Organizations concerned with the ethics of animals include the following:

People for the Ethical Treatment of Animals: https://www.peta.org

The Abolitionist Approach: https://www.abolitionistapproach.com

The Charter for Animal Compassion: https://charterforanimalcompassion.com

World Animal Protection: https://www.worldanimalprotection.org

Organizations concerned with the ethics of machine intelligence include the following:

Cambridge University Institute for Existential Risk: https://www.cser.ac.uk

Center for Human-Compatible Artificial Intelligence: https://humancompatible.ai

DeepMind Safety and Ethics: https://deepmind.com/safety-and-ethics

Leverhulme Centre for the Future of Intelligence: http://lcfi.ac.uk

Machine Intelligence Research Institute: https://intelligence.org

MILA: https://mila.quebec/en

Montreal AI Ethics Institute: https://montrealethics.ai

New York University AI Now Institute: https://ainowinstitute.org

OpenAI: https://openai.com

Oxford University Future of Humanity Institute: https://www.fhi.ox.ac.uk

Stanford University Institute for Human-Centered Artificial Intelligence: https://hai.stanford.edu

The Future of Life Institute: https://futureoflife.org

University of Guelph Artificial Intelligence Centre: https://www.care-ai.ca

Since 2017, many organizations have produced lists of principles concerning the ethics of artificial intelligence, including the following:

1. https://www.ibm.com/watson/assets/duo/pdf/everydayethics.pdf
2. https://standards.ieee.org/content/dam/ieee-standards/standards/web/documents/other/ead1e_general_principles.pdf
3. https://ai.google/principles
4. https://ec.europa.eu/digital-single-market/en/news/draft-ethics-guidelines-trustworthy-ai
5. https://humancompatible.ai
6. https://www.tandfonline.com/doi/pdf/10.1080/09540091.2016.1271400
7. https://futureoflife.org/ai-principles
8. https://www.montrealdeclaration-responsibleai.com/the-declaration
9. https://www.gov.uk/guidance/understanding-artificial-intelligence-ethics-and-safety

In chapter 8, I provide a way of organizing and evaluating these principles.

For a fuller account of ethics with many references, see Thagard, *Natural Philosophy*, chaps. 6–7. R. M. Ryan and E. L. Deci develop a full theory of human psychological needs in *Self-Determination Theory: Basic Psychological*

Needs in Motivation, Development, and Wellness (New York: Guilford, 2017). Computing satisfaction of needs is not a simple calculation but requires coherence algorithms for parallel constraint satisfaction, as described in Thagard, *Natural Philosophy*; and P. Thagard, *Coherence in Thought and Action* (Cambridge, MA: MIT Press, 2000).

L. Gruen, "The Moral Status of Animals," *Stanford Encyclopedia of Psychology* (2017), reviews the moral status of animals. C. M. Korsgaard, *Fellow Creatures: Our Obligations to the Other Animals* (Oxford: Oxford University Press, 2018), argues that animals should be treated as ends in themselves, but says little about how we decide what their ends are, a question best answered in terms of needs. P. Singer, *Animal Liberation: The Definitive Classic of the Animal Movement* (New York: Harper, 2009), takes a utilitarian approach. G. L. Francione and A. Charlton, *Animal Rights: The Abolitionist Approach* (Logan, UT: Exempla Press, 2015), advocates a strong position on animal rights.

Animal morality is defended in M. Bekoff and J. Pierce, *Wild Justice: The Moral Lives of Animals* (Chicago: University of Chicago Press, 2009); De Waal, *Mama's Last Hug*; and S. Vincent, R. Ring, and K. Andrews, "Normative Practices of Other Animals," in *The Routledge Handbook of Moral Epistemology*, ed. K. Jones, M. Timmons, and A. Zimmerman (New York: Routledge, 2019), 57–83. Tomasello, *Becoming Human*, is skeptical. Darwin, *The Descent of Man*, discusses the moral sense of animals. Thagard, "Darwin and the Golden Rule," argues that recursive principles such as the Golden Rule and Kant's categorical imperative are beyond animal understanding. The Golden Rule is found in many religions and cultures; see Wikipedia, https://en.wikipedia.org/wiki/Golden_Rule. J. Zijlmans, R. Marhe, F. Bevaart, M.-J. A. Luijks, L. van Duin, H. Tiemeier, and A. Popma document that moral evaluation uses brain areas involved in emotion in "Neural Correlates of Moral Evaluation and Psychopathic Traits in Male Multi-problem Young Adults," *Frontiers in Psychiatry* 9 (2018): 248.

In *Brain-Mind* and *Natural Philosophy*, I defend a neurocomputational theory of consciousness and provide many references to alternative theories. Substrate independence might provide an argument for moral concern for machines, but Thagard, "Energy Requirements Undermine Substrate Independence and Mind-Body Functionalism" (forthcoming in *Philosophy of Science*), refutes it.

Discussions of AI ethics include O. Bendel, "Considerations about the Relationship between Animal and Machine Ethics," *AI and Society* 31, no.

1 (2016): 103–108; M. Boden, J. Bryson, D. Caldwell, K. Dautenhahn, L. Edwards, S. Kember, P. Newman, et al., "Principles of Robotics: Regulating Robots in the Real World," *Connection Science* 29, no. 2 (2017): 124–129; Boddington, *Towards a Code of Ethics for Artificial Intelligence* (Berlin: Springer, 2017); N. Bostrom and E. Yudkowsky, "The Ethics of Artificial Intelligence," in *The Cambridge Handbook of Artificial Intelligence*, ed. W. Ramsey and K. Frankish (Cambridge: Cambridge University Press, 2014), 316–334; M. Coeckelbergh, *AI Ethics* (Cambridge, MA: MIT Press, 2020); A. Etzioni and O. Etzioni, "AI Assisted Ethics," *Ethics and Information Technology* 18, no. 2 (2016): 149–156; P. Lin, K. Abney, and R. Jenkins, *Robot Ethics 2.0: From Autonomous Cars to Artificial Intelligence* (Oxford: Oxford University Press, 2017); S. Russell, D. Dewey, and M. Tegmark, "Research Priorities for Robust and Beneficial Artificial Intelligence," *AI Magazine* 36, no. 4 (2015): 105–114; "The State of AI Ethics Report (June 2020)," MAIEI, June 24, 2020, https://montrealethics.ai/the-state-of-ai-ethics-report-june-2020. D. Lenat, "EURISKO: A Program That Learns New Heuristics and Domain Concepts," *Artificial Intelligence* 21 (1983): 61–98, describes a program that learns heuristics. The LeCun quote on testosterone comes from Ford, *Architects of Intelligence*, 134. M. Boden argues that machines lack needs in "Robot Says: Whatever," *Aeon* (2018), https://aeon.co/essays/the-robots-wont-take-over-because-they-couldnt-care-less.

The notes for chapter 3 listed surveys about when human-level AI might be achieved. The Kurzweil and Brooks predictions appear in Ford, *Architects of Intelligence*. N. Bostrom considers various scenarios for the development of superintelligence in *Superintelligence: Paths, Dangers, Strategies* (Oxford: Oxford University Press, 2014) but says little about human intelligence or the current state of AI; see also N. Bostrom, A. Dafoe, and C. Flynn, "Policy Desiderata for Superintelligent AI: A Vector Field Approach," in *Ethics of Artificial Intelligence*, ed. S. M. Liao (Oxford: Oxford University Press, 2020).

Thagard, *Natural Philosophy*, advocates empathy for future generations. On animal extinctions, see G. Ceballos, P. R. Ehrlich, and R. Dirzo, "Biological Annihilation via the Ongoing Sixth Mass Extinction Signaled by Vertebrate Population Losses and Declines," *Proceedings of the National Academy of Sciences* 114, no. 30 (2017): E6089–E6096; S. Pimm, P. Raven, A. Peterson, C. H. Sekercioglu, and P. R. Ehrlich, "Human Impacts on the Rates of Recent, Present, and Future Bird Extinctions," *Proceedings of the National Academy of Sciences* 103, no. 29 (2006): 10941–10946; "Media Release: Nature's Dangerous

Decline 'Unprecedented'; Species Extinction Rates 'Accelerating,'" IPBES, n.d., https://www.ipbes.net/news/Media-Release-Global-Assessment. Deep learning contributes to global warming because it requires enormous computer use (Emma Strubell, Ananya Ganesh, and Andrew McCallum, "Energy and Policy Considerations for Deep Learning in NLP," *arXiv preprint arXiv:1906.02243*, 2019). T. Hanson discusses the dangers of bee extinction in *Buzz: The Nature and Necessity of Bees* (New York: Basic Books, 2018), 188. For the number of dogs in the world, see S. Coren, "How Many Dogs Are There in the World?" *Psychology Today*, September 19, 2012, https://www.psychologytoday.com/ca/blog/canine-corner/201209/how-many-dogs-are-there-in-the-world. Octopus numbers are summarized in A. Arkhipkin, "Here's Why Octopus and Squid Populations Are Booming," *New Republic*, May 25, 2016, https://newrepublic.com/article/133734/heres-octopus-squid-populations-booming.

J. Foer describes the brutality of factory farming in *Eating Animals* (New York: Little, Brown, 2009). V. Melina, W. Craig, and S. Levin review the health benefits of vegetarian diets in "Position of the Academy of Nutrition and Dietetics: Vegetarian Diets," *Journal of the Academy of Nutrition and Dietetics* 116, no. 12 (2016): 1970–1980. M. M. Rojas-Downing, A. P. Nejadhashemi, T. Harrigan, and S. A. Woznicki describe the impact of livestock on climate change in "Climate Change and Livestock: Impacts, Adaptation, and Mitigation," *Climate Risk Management* 16 (2017): 145–163. For an argument that keeping pets is unethical, see https://aeon.co/essays/why-keeping-a-pet-is-fundamentally-unethical. D. Benatar, *Better Never to Have Been Born* (Oxford: Clarendon Press, 2006), claims that people are better never to have been born, but is refuted in P. Thagard, *The Brain and the Meaning of Life* (Princeton, NJ: Princeton University Press, 2010). For evidence that most people in prosperous countries are happy, see "World Happiness Report 2018," https://worldhappiness.report/ed/2018; and "Gallup Global Well-Being," https://news.gallup.com/poll/126965/gallup-global-wellbeing.aspx.

D. Acemoglu and P. Restrepo review the employment effects of robots in "Robots and Jobs: Evidence from US Labor Markets," *NBER Working Paper* (w23285). S. Zuboff, *The Age of Surveillance Capitalism: The Fight for a Human Future at the New Frontier of Power* (New York: PublicAffairs, 2019), describes how technology produces surveillance capitalism. A. Caliskan, J. J. Bryson, and A. Narayanan find humanlike biases in the results of machine learning in "Semantics Derived Automatically from Language Corpora Contain Human-Like Biases," *Science* 356, no. 6334 (2017): 183–186. Many researchers and companies have made a pledge to ban lethal autonomous weapons: https://futureoflife.org/lethal-autonomous-weapons-pledge. S. Vallor, *Technology and the Virtues: A Philosophical Guide to a Future Worth Wanting* (New York: Oxford University Press, 2016), takes a virtue ethics approach to killer robots, but I think human needs are more relevant than virtues such as courage. Thagard, *Natural Philosophy*, argues that values are emotional attitudes that can nevertheless be objective. The Gandhi quote on ethics comes from https://en.wikiquote.org/wiki/Need.

Chapter 8: The Ethics of Artificial Intelligence

A. Jobin, M. Ienca, and E. Vayena cover more than eighty lists of principles for ethical AI in "The Global Landscape of AI Ethics Guidelines," *Nature Machine Intelligence* 1, no. 9 (2019): 389–399; the quote comes from p. 389. S. Russell claims that beneficial AI concerns people's preferences rather than their ethical principles and values in *Human Compatible: Artificial Intelligence and the Problem of Control* (New York: Penguin Random House, 2019).

T. L. Beauchamp and J. F. Childress advocate the four principles for medical ethics in *Principles of Biomedical Ethics* (New York: Oxford University Press, 2013). The United Nations principles on prisoners are at https://www.ohchr.org/EN/ProfessionalInterest/Pages/MedicalEthics.aspx.

On the Asilomar AI Principles, see https://futureoflife.org/ai-principles; and Tegmark, *Life 3.0*.

Psychological alternatives to traditional economic explanations include D. Kahneman, *Thinking Fast and Slow* (Toronto: Doubleday, 2011); and Thagard, *Brain-Mind*, chap. 7.

Preference utilitarianism is advocated in R. M. Hare, *Moral Thinking: Its Levels, Methods, and Point* (Oxford: Oxford University Press, 1981), and critiqued in D. Hausman, M. McPherson, and D. Satz, *Economic Analysis, Moral Philosophy, and Public Policy* (Cambridge: Cambridge University Press, 2017), 128–129.

On how concepts generally have associated emotional values, see E. Cambria, "Affective Computing and Sentiment Analysis," *IEEE Intelligence Systems* 31, no. 2 (2016): 102–107; R. H. Fazio, "On the Automatic Activation of Associated Evaluations: An Overview," *Cognition and Emotion* 15 (2001): 115–141; and Thagard, *Mind-Society*. D. Keltner, K. Oatley, and J. M. Jenkins review emotions in *Understanding Emotions* (New York: Wiley,

2018). AI values are discussed by B. Christian, *The Alignment Problem: Machine Learning and Human Values* (New York: Penguin, 2020).

Concepts combine exemplars, stereotypes, and explanations: Blouw et al., "Concepts as Semantic Pointers"; Murphy, *The Big Book of Concepts*; Thagard, *Brain-Mind*. The three-analysis of the concept *value* comes from Thagard, *Natural Philosophy*, 151. Project: do three-analyses of important AI values such as transparency and privacy.

Brandeis and Warren defend a right to privacy in "The Right to Privacy," *Harvard Law Review* 4 (1890): 193–220.

Applications of cognitive-affective maps include T. Homer-Dixon, M. Milkoreit, S. J. Mock, T. Schröder, and P. Thagard, "The Conceptual Structure of Social Disputes: Cognitive-Affective Maps as a Tool for Conflict Analysis and Resolution," *SAGE Open* 4 (2014); P. Thagard, "EMPATHICA: A Computer Support System with Visual Representations for Cognitive-Affective Mapping," in *Proceedings of the Workshop on Visual Reasoning and Representation*, ed. K. McGregor (Menlo Park, CA: AAAI Press, 2010), 79–81; and Thagard, *Mind-Society*.

Production rule systems include GPS (Newell and Simon, *Human Problem Solving*), ACT-R (Anderson, *Rules of the Mind*), and SOAR (John E. Laird, *The Soar Cognitive Architecture* [Cambridge, MA: MIT Press, 2012]). On default rules, see Holland et al., *Induction*. Eliasmith, *How to Build a Brain*, shows how rules can be translated into spiking neurons.

S. Marsella and J. Gratch, in both "Computationally Modeling Human Emotion," *Communications of the ACM* 57, no. 12 (2014): 56–67, and "Computational Models of Emotions as Psychological Tools," in *Handbook of Emotions*, ed. L. F. Barrett, M. Lewis, and J. M. Haviland-Jones (New York: Guilford, 2016), 113–129, review computational models of emotions, which involve physiological changes and conscious feelings as well as appraisals: Damasio, *Descartes' Error: Emotion, Reason, and the Human Brain* (New York: G. P. Putnam's Sons, 1994); Kajić et al., "Semantic Pointer Theory of Emotions," 35–53.

Minsky, "A Framework for Representing Knowledge," 211–277, presents frames.

The theory of emotional coherence comes from P. Thagard, *Hot Thought: Mechanisms and Application of Emotional Cognition* (Cambridge, MA: MIT Press, 2006); and coherence algorithms are analyzed in P. Thagard and K. Verbeurgt, "Coherence as Constraint Satisfaction," *Cognitive Science* 22 (1998): 1–24.

P. Cudney discusses casuistry in ethics in "What Really Separates Casuistry from Principilism in Biomedical Ethics," *Theoretical Medicine in Bioethics* 35, no. 3 (2014): 205–229. Algorithms for analogical inference and case-based reasoning include Forbus et al., "Extending SME to Handle Larger-Scale Cognitive Modeling"; Holyoak and Thagard, *Mental Leaps*; and Leake, *Case-Based Reasoning*.

Index

Page numbers followed by a "b," "f," or "t" indicate boxes, figures, or tables, respectively.